Yamaha T50 & 80 Townmate Owners Workshop Manual

by Pete Shoemark

Models covered
T50 Townmate. 49cc. May 1986 to February 1989
T80 Townmate. 79cc. June 1983 on

(1247-3U4)

Haynes Group Limited
Haynes North America, Inc

www.haynes.com

Acknowledgements

Our thanks are due to Jim Patch of Yeovil Motorcycle Services who supplied the machine featured throughout this manual; Mitsui Machinery Sales (UK) Ltd, who provided the necessary service information and who gave permission to reproduce many of the line drawings used; and the staff of Mitsui's Technical Support Group for their technical advice and assistance.

We would also like to thank the Avon Rubber Company who kindly supplied information and technical assistance on tyre fitting, and NGK Spark Plugs (UK) Ltd for information on spark plug maintenance and electrode conditions.

A book in the **Haynes Owners Workshop Manual Series**

ISBN 978 1 85960 068 9

British Library Cataloguing in Publication Data
A catalogue record for this book is available from the British Library

Contents

Left-hand view of the T80

Engine/transmission unit

About this manual

The purpose of this manual is to present the owner with a concise and graphic guide which will enable him to tackle any operation from basic routine maintenance to a major overhaul. It has been assumed that any work would be undertaken without the luxury of a well-equipped workshop and a range of manufacturer's service tools.

To this end, the machine featured in the manual was stripped and rebuilt in our own workshop, by a team comprising a mechanic, a photographer and the author. The resulting photographic sequence depicts events as they took place, the hands shown being those of the author and the mechanic.

The use of specialised, and expensive, service tools was avoided unless their use was considered to be essential due to risk of breakage or injury. There is usually some way of improvising a method of removing a stubborn component, providing that a suitable degree of care is exercised.

The author learnt his motorcycle mechanics over a number of years, faced with the same difficulties and using similar facilities to those encountered by most owners. It is hoped that this practical experience can be passed on through the pages of this manual.

Where possible, a well-used example of the machine is chosen for the workshop project, as this highlights any areas which might be particularly prone to giving rise to problems. In this way, any such difficulties are encountered and resolved before the text is written, and the techniques used to deal with them can be incorporated in the relevant section. Armed with a working knowledge of the machine, the author undertakes a considerable amount of research in order that the maximum amount of data can be included in the manual.

A comprehensive section, preceding the main part of the manual, describes procedures for carrying out the routine maintenance of the machine at intervals of time and mileage. This section is included particularly for those owners who wish to ensure the efficient day-to-day running of their motorcycle, but who choose not to undertake overhaul or renovation work.

Each Chapter is divided into numbered sections. Within these sections are numbered paragraphs. Cross reference throughout the manual is quite straightforward and logical. When reference is made 'See Section 6.10' it means Section 6, paragraph 10 in the same Chapter. If another Chapter were intended, the reference would read, for example, 'See Chapter 2, Section 6.10'. All the photographs are captioned with a section/paragraph number to which they refer and are relevant to the Chapter text adjacent.

Figures (usually line illustrations) appear in a logical but numerical order, within a given Chapter. Fig. 1.1 therefore refers to the first figure in Chapter 1.

Left-hand and right-hand descriptions of the machines and their components refer to the left and right of a given machine when the rider is seated normally.

Motorcycle manufacturers continually make changes to specifications and recommendations, and these, when notified, are incorporated into our manuals at the earliest opportunity.

Introduction to the Yamaha T50 and 80 models

June 1983 saw the introduction of the first Townmate model, the T80. The T80 received minor modification in May 1986, at the same time as the smaller-capacity T50 was introduced.

The machine follows the now well-established 'step-through' design familiar in almost every country in the world, being based around an open spine frame with the engine/transmission unit positioned below the tubular member running between the main spine section and the steering head. The fuel tank is located within the frame main section, below the seat, leaving the area forward of the seat nose open. The engine/transmission unit is partially hidden by moulded plastic legshields which also serve to provide good weather protection for the rider.

Front suspension is by leading link forks, the wheel being carried by oil-damped telescopic suspension units. Rear suspension is by swinging arm, supported on conventional oil-damped telescopic units. The wheels are of relatively large diameter, which contribute much to the machine's superior stability when compared with the fashionable scooter-inspired mopeds and small motorcycles.

The engine is a simple single overhead camshaft, single cylinder four-stroke unit, the cylinder being angled forward to allow the unit to be accommodated beneath the frame. Transmission is by a foot-operated three-speed (T50) or four-speed (T80) gearbox working in conjunction with a centrifugal clutch. The latter arrangement dispenses with the usual manual clutch lever; the clutch engages automatically as the engine speed is increased. To facilitate gear changing, the clutch is linked to the gearchange pedal. This disengages the clutch as the pedal is depressed or raised at each change, drive being resumed when the pedal is released.

The T80 is notable in its use of shaft final drive in place of the conventional enclosed chain more commonly used on this type of machine. Whilst shaft drive is certainly heavier than an equivalent chain and sprocket arrangement, it removes the need for periodic adjustment of the rear wheel to compensate for chain wear and will normally require no attention from the rider apart from renewal of its lubricant.

Early and late T80 models are identified in the text by their model code numbers, where necessary. Refer to the following identification details for information:

Model	Code no	Engine/frame no
T50	2FM	2FM-000101 on
T80	35T	35T-000101 on
T80	2FL	35T-030101 on

Dimensions and weights

Overall length	1860 mm (73.2 in)
Overall width	670 mm (26.4 in)
Overall height	1050 mm (41.3 in)
Wheelbase	1180 mm (46.5 in)
Seat height	750 mm (29.5 in)
Ground clearance	130 mm (5.1 in)
Weight (with oil and full fuel tank)	87 kg (192 lb)

Ordering spare parts

When replacement parts are required for your Yamaha, it is advisable to deal with an authorized Yamaha dealer. He is in the best position to offer specialist advice and will be able to supply the more commonly used parts from stock. If the parts need to be ordered, remember that an official dealer will be able to arrange faster delivery than a non-specialist supplier. Try to order parts well in advance where this is possible. For example, read through the appropriate section of the manual and see whether gaskets or seals will be needed. This can often avoid having the machine off the road for a week or two while they are ordered.

When ordering, always quote the machine details in full. This will ensure that the correct parts are supplied and will take into account any retrospective manufacturer's modifications. You will need the frame number, which is stamped into the frame, below the left-hand side panel, and the engine number, which is stamped into the top of the left-hand engine casing.

During the initial warranty period, and as a general rule, make sure that only genuine Yamaha parts are used. Fitting non-standard parts may well invalidate the warranty, and more importantly, could prove dangerous. Be particularly wary of pattern safety-related-parts such as brake and suspension components. These often resemble the original parts very closely and may even be supplied in counterfeit packaging and sold as genuine items.

Some of the more consumable items, such as spark plugs, bulbs, oils, greases and tyres can be purchased from local sources like accessory shops and motor factors, or from mail order suppliers. Always stick to well-known and reputable brands and make sure that the items supplied are suitable to your machine. When buying tyres, be warned that some perfectly good makes of tyre may not be suited to the suspension characteristics of your model. An authorized Yamaha dealer or any good tyre supplier will be able to advise here; if they seem vague or non-commital, go elsewhere.

Engine number is stamped into the crankcase below left-hand rear edge of the legshield

Frame number is stamped into main frame section, below left-hand side panel

Safety first!

Professional motor mechanics are trained in safe working procedures. However enthusiastic you may be about getting on with the job in hand, do take the time to ensure that your safety is not put at risk. A moment's lack of attention can result in an accident, as can failure to observe certain elementary precautions.

There will always be new ways of having accidents, and the following points do not pretend to be a comprehensive list of all dangers; they are intended rather to make you aware of the risks and to encourage a safety-conscious approach to all work you carry out on your vehicle.

Essential DOs and DON'Ts

DON'T start the engine without first ascertaining that the transmission is in neutral.

DON'T suddenly remove the filler cap from a hot cooling system – cover it with a cloth and release the pressure gradually first, or you may get scalded by escaping coolant.

DON'T attempt to drain oil until you are sure it has cooled sufficiently to avoid scalding you.

DON'T grasp any part of the engine, exhaust or silencer without first ascertaining that it is sufficiently cool to avoid burning you.

DON'T allow brake fluid or antifreeze to contact the machine's paintwork or plastic components.

DON'T syphon toxic liquids such as fuel, brake fluid or antifreeze by mouth, or allow them to remain on your skin.

DON'T inhale dust – it may be injurious to health (see *Asbestos* heading).

DON'T allow any spilt oil or grease to remain on the floor – wipe it up straight away, before someone slips on it.

DON'T use ill-fitting spanners or other tools which may slip and cause injury.

DON'T attempt to lift a heavy component which may be beyond your capability – get assistance.

DON'T rush to finish a job, or take unverified short cuts.

DON'T allow children or animals in or around an unattended vehicle.

DON'T inflate a tyre to a pressure above the recommended maximum. Apart from overstressing the carcase and wheel rim, in extreme cases the tyre may blow off forcibly.

DO ensure that the machine is supported securely at all times. This is especially important when the machine is blocked up to aid wheel or fork removal.

DO take care when attempting to slacken a stubborn nut or bolt. It is generally better to pull on a spanner, rather than push, so that if slippage occurs you fall away from the machine rather than on to it.

DO wear eye protection when using power tools such as drill, sander, bench grinder etc.

DO use a barrier cream on your hands prior to undertaking dirty jobs – it will protect your skin from infection as well as making the dirt easier to remove afterwards; but make sure your hands aren't left slippery. Note that long-term contact with used engine oil can be a health hazard.

DO keep loose clothing (cuffs, tie etc) and long hair well out of the way of moving mechanical parts.

DO remove rings, wristwatch etc, before working on the vehicle – especially the electrical system.

DO keep your work area tidy – it is only too easy to fall over articles left lying around.

DO exercise caution when compressing springs for removal or installation. Ensure that the tension is applied and released in a controlled manner, using suitable tools which preclude the possibility of the spring escaping violently.

DO ensure that any lifting tackle used has a safe working load rating adequate for the job.

DO get someone to check periodically that all is well, when working alone on the vehicle.

DO carry out work in a logical sequence and check that everything is correctly assembled and tightened afterwards.

DO remember that your vehicle's safety affects that of yourself and others. If in doubt on any point, get specialist advice.

IF, in spite of following these precautions, you are unfortunate enough to injure yourself, seek medical attention as soon as possible.

Asbestos

Certain friction, insulating, sealing, and other products – such as brake linings, clutch linings, gaskets, etc – contain asbestos. *Extreme care must be taken to avoid inhalation of dust from such products since it is hazardous to health.* If in doubt, assume that they *do* contain asbestos.

Fire

Remember at all times that petrol (gasoline) is highly flammable. Never smoke, or have any kind of naked flame around, when working on the vehicle. But the risk does not end there – a spark caused by an electrical short-circuit, by two metal surfaces contacting each other, by careless use of tools, or even by static electricity built up in your body under certain conditions, can ignite petrol vapour, which in a confined space is highly explosive.

Always disconnect the battery earth (ground) terminal before working on any part of the fuel or electrical system, and never risk spilling fuel on to a hot engine or exhaust.

It is recommended that a fire extinguisher of a type suitable for fuel and electrical fires is kept handy in the garage or workplace at all times. Never try to extinguish a fuel or electrical fire with water.

Note: *Any reference to a 'torch' appearing in this manual should always be taken to mean a hand-held battery-operated electric lamp or flashlight. It does* **not** *mean a welding/gas torch or blowlamp.*

Fumes

Certain fumes are highly toxic and can quickly cause unconsciousness and even death if inhaled to any extent. Petrol (gasoline) vapour comes into this category, as do the vapours from certain solvents such as trichloroethylene. Any draining or pouring of such volatile fluids should be done in a well ventilated area.

When using cleaning fluids and solvents, read the instructions carefully. Never use materials from unmarked containers – they may give off poisonous vapours.

Never run the engine of a motor vehicle in an enclosed space such as a garage. Exhaust fumes contain carbon monoxide which is extremely poisonous; if you need to run the engine, always do so in the open air or at least have the rear of the vehicle outside the workplace.

The battery

Never cause a spark, or allow a naked light, near the vehicle's battery. It will normally be giving off a certain amount of hydrogen gas, which is highly explosive.

Always disconnect the battery earth (ground) terminal before working on the fuel or electrical systems.

If possible, loosen the filler plugs or cover when charging the battery from an external source. Do not charge at an excessive rate or the battery may burst.

Take care when topping up and when carrying the battery. The acid electrolyte, even when diluted, is very corrosive and should not be allowed to contact the eyes or skin.

If you ever need to prepare electrolyte yourself, always add the acid slowly to the water, and never the other way round. Protect against splashes by wearing rubber gloves and goggles.

Mains electricity and electrical equipment

When using an electric power tool, inspection light etc, always ensure that the appliance is correctly connected to its plug and that, where necessary, it is properly earthed (grounded). Do not use such appliances in damp conditions and, again, beware of creating a spark or applying excessive heat in the vicinity of fuel or fuel vapour. Also ensure that the appliances meet the relevant national safety standards.

Ignition HT voltage

A severe electric shock can result from touching certain parts of the ignition system, such as the HT leads, when the engine is running or being cranked, particularly if components are damp or the insulation is defective. Where an electronic ignition system is fitted, the HT voltage is much higher and could prove fatal.

Tools and working facilities

The first priority when undertaking maintenance or repair work of any sort on a motorcycle is to have a clean, dry, well-lit working area. Work carried out in peace and quiet in the well-ordered atmosphere of a good workshop will give more satisfaction and much better results than can usually be achieved in poor working conditions. A good workshop must have a clean flat workbench or a solidly constructed table of convenient working height. The workbench or table should be equipped with a vice which has a jaw opening of at least 4 in (100 mm). A set of jaw covers should be made from soft metal such as aluminium alloy or copper, or from wood. These covers will minimise the marking or damaging of soft or delicate components which may be clamped in the vice. Some clean, dry, storage space will be required for tools, lubricants and dismantled components. It will be necessary during a major overhaul to lay out engine/gearbox components for examination and to keep them where they will remain undisturbed for as long as is necessary. To this end it is recommended that a supply of metal or plastic containers of suitable size is collected. A supply of clean, lint-free, rags for cleaning purposes and some newspapers, other rags, or paper towels for mopping up spillages should also be kept. If working on a hard concrete floor note that both the floor and one's knees can be protected from oil spillages and wear by cutting open a large cardboard box and spreading it flat on the floor under the machine or workbench. This also helps to provide some warmth in winter and to prevent the loss of nuts, washers, and other tiny components which have a tendency to disappear when dropped on anything other than a perfectly clean, flat, surface.

Unfortunately, such working conditions are not always available to the home mechanic. When working in poor conditions it is essential to take extra time and care to ensure that the components being worked on are kept scrupulously clean and to ensure that no components or tools are lost or damaged.

A selection of good tools is a fundamental requirement for anyone contemplating the maintenance and repair of a motor vehicle. For the owner who does not possess any, their purchase will prove a considerable expense, offsetting some of the savings made by doing-it-yourself. However, provided that the tools purchased meet the relevant national safety standards and are of good quality, they will last for many years and prove an extremely worthwhile investment.

To help the average owner to decide which tools are needed to carry out the various tasks detailed in this manual, we have compiled three lists of tools under the following headings: *Maintenance and minor repair, Repair and overhaul*, and *Specialized*. The newcomer to practical mechanics should start off with the simpler jobs around the vehicle. Then, as his confidence and experience grow, he can undertake more difficult tasks, buying extra tools as and when they are needed. In this way, a *Maintenance and minor repair* tool kit can be built-up into a *Repair and overhaul* tool kit over a considerable period of time without any major cash outlays. The experienced home mechanic will have a tool kit good enough for most repair and overhaul procedures and will add tools from the specialized category when he feels the expense is justified by the amount of use these tools will be put to.

It is obviously not possible to cover the subject of tools fully here. For those who wish to learn more about tools and their use there is a book entitled *Motorcycle Workshop Practice Manual* (Book No 1454) available from the publishers of this manual.

As a general rule, it is better to buy the more expensive, good quality tools. Given reasonable use, such tools will last for a very long time, whereas the cheaper, poor quality, item will wear out faster and need to be renewed more often, thus nullifying the original saving. There is also the risk of a poor quality tool breaking while in use, causing personal injury or expensive damage to the component being worked on.

For practically all tools, a tool factor is the best source since he will have a very comprehensive range compared with the average garage or accessory shop. Having said that, accessory shops often offer excellent quality tools at discount prices, so it pays to shop around. There are plenty of tools around at reasonable prices, but always aim to purchase items which meet the relevant national safety standards. If in doubt, seek the advice of the shop proprietor or manager before making a purchase.

The basis of any toolkit is a set of spanners. While open-ended spanners with their slim jaws, are useful for working on awkwardly-positioned nuts, ring spanners have advantages in that they grip the nut far more positively. There is less risk of the spanner slipping off the nut and damaging it, for this reason alone ring spanners are to be preferred. Ideally, the home mechanic should acquire a set of each, but if expense rules this out a set of combination spanners (open-ended at one end and with a ring of the same size at the other) will provide a good compromise. Another item which is so useful it should be considered an essential requirement for any home mechanic is a set of socket spanners. These are available in a variety of drive sizes. It is recommended that the ½-inch drive type is purchased to begin with as although bulkier and more expensive than the ⅜-inch type, the larger size is far more common and will accept a greater variety of torque wrenches, extension pieces and socket sizes. The socket set should comprise sockets of sizes between 8 and 24 mm, a reversible ratchet drive, an extension bar of about 10 inches in length, a spark plug socket with a rubber insert, and a universal joint. Other attachments can be added to the set at a later date.

Maintenance and minor repair tool kit

Set of spanners 8 – 24 mm
Set of sockets and attachments
Spark plug spanner with rubber insert – 10, 12, or 14 mm as appropriate
Adjustable spanner
C-spanner/pin spanner
Torque wrench (same size drive as sockets)
Set of screwdrivers (flat blade)
Set of screwdrivers (cross-head)
Set of Allen keys 4 – 10 mm
Impact screwdriver and bits
Ball pein hammer – 2 lb
Hacksaw (junior)
Self-locking pliers – Mole grips or vice grips
Pliers – combination
Pliers – needle nose
Wire brush (small)
Soft-bristled brush
Tyre pump
Tyre pressure gauge
Tyre tread depth gauge
Oil can
Fine emery cloth
Funnel (medium size)
Drip tray
Grease gun
Set of feeler gauges
Strobe timing light
Continuity tester (dry battery and bulb)
Soldering iron and solder
Wire stripper or craft knife
PVC insulating tape
Assortment of split pins, nuts, bolts, and washers

Repair and overhaul toolkit

The tools in this list are virtually essential for anyone undertaking major repairs to a motorcycle and are additional to the tools listed above. Concerning Torx driver bits, Torx screws are encountered on some of the more modern machines where their use is restricted to fastening certain components inside the engine/gearbox unit. It is therefore recommended that if Torx bits cannot be borrowed from a local dealer, they are purchased individually as the need arises. They are not in regular use in the motor trade and will therefore only be available in specialist tool shops.

Plastic or rubber soft-faced mallet
Torx driver bits
Pliers – electrician's side cutters
Circlip pliers – internal (straight or right-angled tips are available)
Circlip pliers – external
Cold chisel
Centre punch
Pin punch
Scriber
Scraper (made from soft metal such as aluminium or copper)
Soft metal drift
Steel rule/straight edge
Assortment of files
Electric drill and bits
Wire brush (large)
Soft wire brush (similar to those used for cleaning suede shoes)
Sheet of plate glass
Hacksaw (large)
Valve grinding tool
Valve grinding compound (coarse and fine)
Stud extractor set (E-Z out)

Specialized tools

This is not a list of the tools made by the machine's manufacturer to carry out a specific task on a limited range of models. Occasional references are made to such tools in the text of this manual and, in general, an alternative method of carrying out the task without the manufacturer's tool is given where possible. The tools mentioned in this list are those which are not used regularly and are expensive to buy in view of their infrequent use. Where this is the case it may be possible to hire or borrow the tools against a deposit from a local dealer or tool hire shop. An alternative is for a group of friends or a motorcycle club to join in the purchase.

Valve spring compressor
Piston ring compressor
Universal bearing puller
Cylinder bore honing attachment (for electric drill)
Micrometer set
Vernier calipers
Dial gauge set
Cylinder compression gauge
Multimeter
Dwell meter/tachometer

Care and maintenance of tools

Whatever the quality of the tools purchased, they will last much longer if cared for. This means in practice ensuring that a tool is used for its intended purpose; for example screwdrivers should not be used as a substitute for a centre punch, or as chisels. Always remove dirt or grease and any metal particles but remember that a light film of oil will prevent rusting if the tools are infrequently used. The common tools can be kept together in a large box or tray but the more delicate, and more expensive, items should be stored separately where they cannot be damaged. When a tool is damaged or worn out, be sure to renew it immediately. It is false economy to continue to use a worn spanner or screwdriver which may slip and cause expensive damage to the component being worked on.

Fastening systems

Fasteners, basically, are nuts, bolts and screws used to hold two or more parts together. There are a few things to keep in mind when working with fasteners. Almost all of them use a locking device of some type; either a lock washer, lock nut, locking tab or thread adhesive. All threaded fasteners should be clean, straight, have undamaged threads and undamaged corners on the hexagon head where the spanner fits. Develop the habit of replacing all damaged nuts and bolts with new ones.

Rusted nuts and bolts should be treated with a rust penetrating fluid to ease removal and prevent breakage. After applying the rust penetrant, let it 'work' for a few minutes before trying to loosen the nut or bolt. Badly rusted fasteners may have to be chiseled off or removed with a special nut breaker, available at tool shops.

Flat washers and lock washers, when removed from an assembly should always be replaced exactly as removed. Replace any damaged washers with new ones. Always use a flat washer between a lock washer and any soft metal surface (such as aluminium), thin sheet metal or plastic. Special lock nuts can only be used once or twice before they lose their locking ability and must be renewed.

If a bolt or stud breaks off in an assembly, it can be drilled out and removed with a special tool called an E-Z out. Most dealer service departments and motorcycle repair shops can perform this task, as well as others (such as the repair of threaded holes that have been stripped out).

Spanner size comparison

Jaw gap (in)	Spanner size
0.250	1/4 in AF
0.276	7 mm
0.313	5/16 in AF
0.315	8 mm
0.344	11/32 in AF; 1/8 in Whitworth
0.354	9 mm
0.375	3/8 in AF
0.394	10 mm
0.433	11 mm
0.438	7/16 in AF
0.445	3/16 in Whitworth; 1/4
0.472	12 mm
0.500	1/2 in AF
0.512	13 mm
0.525	1/4 in Whitworth; 5/16 in BSF
0.551	14 mm
0.563	9/16 in AF
0.591	15 mm
0.600	5/16 in Whitworth; 3/8 in BSF
0.625	5/8 in AF
0.630	16 mm
0.669	17 mm
0.686	11/16 in AF
0.709	18 mm
0.710	3/8 in Whitworth; 7/16 in BSF
0.748	19 mm
0.750	3/4 in AF
0.813	13/16 in AF
0.820	7/16 in Whitworth; 1/2 in BSF
0.866	22 mm
0.875	7/8 in AF
0.920	1/2 in Whitworth; 9/16 in BSF
0.938	15/16 in AF

Jaw gap (in)	Spanner size
0.945	24 mm
1.000	1 in AF
1.010	9/16 in Whitworth; 5/8 in BSF
1.024	26 mm
1.063	1 1/16 in AF; 27 mm
1.100	5/16 in Whitworth; 11/16 in BSF
1.125	1 1/8 in AF
1.181	30 mm
1.200	11/16 in Whitworth; 3/4 in BSF
1.250	1 1/4 in AF
1.260	32 mm
1.300	3/4 in Whitworth; 7/8 in BSF
1.313	1 5/16 in AF
1.390	13/16 in Whitworth; 15/16 in BSF
1.417	36 mm
1.438	1 7/16 in AF
1.480	7/8 in Whitworth; 1 in BSF
1.500	1 1/2 in AF
1.575	40 mm; 15/16 in Whitworth
1.614	41 mm
1.625	1 5/8 in AF
1.670	1 in Whitworth; 1 1/8 in BSF
1.688	1 11/16 in AF
1.811	46 mm
1.813	1 13/16 in AF
1.860	1 1/8 in Whitworth; 1 1/4 in BSF
1.875	1 7/8 in AF
1.969	50 mm
2.000	2 in AF
2.050	1 1/4 in Whitworth; 1 3/8 in BSF
2.165	55 mm
2.362	60 mm

Standard torque settings

Specific torque settings will be found at the end of the specifications section of each chapter. Where no figure is given, bolts should be secured according to the table below.

Fastener type (thread diameter)	**kgf m**	**lbf ft**
5mm bolt or nut	0.45 – 0.6	3.5 – 4.5
6 mm bolt or nut	0.8 – 1.2	6 – 9
8 mm bolt or nut	1.8 – 2.5	13 – 18
10 mm bolt or nut	3.0 – 4.0	22 – 29
12 mm bolt or nut	5.0 – 6.0	36 – 43
5 mm screw	0.35 – 0.5	2.5 – 3.6
6 mm screw	0.7 – 1.1	5 – 8
6 mm flange bolt	1.0 – 1.4	7 – 10
8 mm flange bolt	2.4 – 3.0	17 – 22
10 mm flange bolt	3.5 – 4.5	25 – 33

Choosing and fitting accessories

The range of accessories available to the modern motorcyclist is almost as varied and bewildering as the range of motorcycles. This Section is intended to help the owner in choosing the correct equipment for his needs and to avoid some of the mistakes made by many riders when adding accessories to their machines. It will be evident that the Section can only cover the subject in the most general terms and so it is recommended that the owner, having decided that he wants to fit, for example, a luggage rack or carrier, seeks the advice of several local dealers and the owners of similar machines. This will give a good idea of what makes of carrier are easily available, and at what price. Talking to other owners will give some insight into the drawbacks or good points of any one make. A walk round the motorcycles in car parks or outside a dealer will often reveal the same sort of information.

The first priority when choosing accessories is to assess exactly what one needs. It is, for example, pointless to buy a large heavy-duty carrier which is designed to take the weight of fully laden panniers and topbox when all you need is a place to strap on a set of waterproofs and a lunchbox when going to work. Many accessory manufacturers have ranges of equipment to cater for the individual needs of different riders and this point should be borne in mind when looking through a dealer's catalogues. Having decided exactly what is required and the use to which the accessories are going to be put, the owner will need a few hints on what to look for when making the final choice. To this end the Section is now sub-divided to cover the more popular accessories fitted. Note that it is in no way a customizing guide, but merely seeks to outline the practical considerations to be taken into account when adding aftermarket equipment to a motorcycle.

It should be noted that the design of the Yamaha T80 models provides some measure of weather protection as standard, in the form of the legshields. This can be supplemented by a windshield, or handlebar fairing, and other useful items like a rack or top box can be fitted. The semi-enclosed nature of the machine rules out the fitting of certain other accessories, as does the limited power available from the electrical system. It is recommended that the advice of an authorized Yamaha dealer is sought before purchasing any accessory part, thus ensuring that the intended purchase is suitable and can be fitted to this model.

Windshields and handlebar fairings

As has been suggested, the weather protection afforded by the legshields can be extended by fitting a suitable windshield to the handlebars. Of the many types available, the clear acrylic sheet type, supported by tubular or round bar stays is probably best. Before purchasing, be absolutely certain that the brackets supplied will allow it to be fitted to the Yamaha T80 models; many so-called universal types cannot easily be fitted to machines with handlebar nacelles.

Fitting should be a relatively straightforward task, and should be carried out by following the installation instructions closely. Alternatively, the dealer should be able to fit the windshield for you for a small fee. The acrylic screen should be positioned so that you can just see over the top when seated normally, not so that you look through it. This may seem an odd recommendation, but in practice it is not easy to see through a plastic screen safely, especially at night, in rain, or if the screen becomes dirty or scratched. Once on the move, the flow of air over the top edge of the screen will deflect rain clear of a helmet visor or goggles.

When using the machine for the first time, be prepared for a somewhat altered feel to the steering due to the effect of wind pressure on the screen; this is not dangerous, but will require a little accustomisation.

Luggage racks or carriers

Carriers are possibly the commonest item to be fitted to modern motorcycles. They vary enormously in size, carrying capacity, and durability. When selecting a carrier, always look for one which is made specifically for your machine and which is bolted on with as few separate brackets as possible. The universal-type carrier, with its mass of brackets and adaptor pieces, will generally prove too weak to be of any real use. A good carrier should bolt to the main frame, generally using the two suspension unit top mountings and a mudguard mounting bolt as attachment points, and have its luggage platform as low and as far forward as possible to minimise the effect of any load on the machine's stability. Look for good quality, heavy gauge tubing, good welding and good finish. Also ensure that the carrier does not prevent opening of the seat, sidepanels or tail compartment, as appropriate. When using a carrier, be very careful not to overload it. Excessive weight placed so high and so far to the rear of any motorcycle will have an adverse effect on the machine's steering and stability.

Luggage

Motorcycle luggage can be grouped under two headings: soft and hard. Both types are available in many sizes and styles and have advantages and disadvantages in use.

Soft luggage is now becoming very popular because of its lower cost and its versatility. Whether in the form of tankbags, panniers, or strap-on bags, soft luggage requires in general no brackets and no modification to the motorcycle. Equipment can be swapped easily from one motorcycle to another and can be fitted and removed in seconds. Awkwardly shaped loads can easily be carried. The disadvantages of soft luggage are that the contents cannot be secure against the casual thief, very little protection is afforded in the event of a crash, and waterproofing is generally poor. Also, in the case of panniers, carrying capacity is restricted to approximately 10 lb, although this amount will vary considerably depending on the manufacturer's recommendation. When purchasing soft luggage, look for good quality material, generally vinyl or nylon, with strong, well-stitched attachment points. It is always useful to have separate pockets, especially on tank bags, for items which will be needed on the journey. When purchasing a tank bag, look for one which has a separate, well-padded, base. This will protect the tank's paintwork and permit easy access to the filler cap at petrol stations.

Hard luggage is confined to two types: panniers, and top boxes or tail trunks. Most hard luggage manufacturers produce matching sets of these items, the basis of which is generally that manufacturer's own heavy-duty luggage rack. Variations on this theme occur in the form of separate frames for the better quality panniers, fixed or quickly-detachable luggage, and in size and carrying capacity. Hard luggage offers a reasonable degree of security against theft and good protection against weather and accident damage. Carrying capacity is greater than that of soft luggage, around 15 – 20 lb in the case of panniers, although top boxes should never be loaded as much as their apparent capacity might imply. A top box should only be used for lightweight items, because one that is heavily laden can have a serious effect on the stability of the machine. When purchasing hard luggage look for the same good points as mentioned under Windshields and handlebar fairings, ie good quality mounting brackets and fittings, and well-finished fibreglass or ABS plastic cases. Again as with fairings, always purchase luggage made specifically for your motorcycle, using as few separate brackets as possible, to ensure that everything remains securely bolted in place. When fitting hard luggage, be careful to check that the rear suspension and brake operation will not be impaired in any way and remember that many pannier kits require re-siting of the indicators. Remember also that a non-standard exhaust system may make fitting extremely difficult.

Accessories – general

Accessories fitted to your motorcycle will rapidly deteriorate if not cared for. Regular washing and polishing will maintain the finish and will provide an opportunity to check that all mounting bolts and nuts are securely fastened. Any signs of chafing or wear should be watched for, and the cause cured as soon as possible before serious damage occurs.

As a general rule, do not expect the re-sale value of your motorcycle to increase by an amount proportional to the amount of money and effort put into fitting accessories. It is usually the case that an absolutely standard motorcycle will sell more easily at a better price than one that has been modified. If you are in the habit of exchanging your machine for another at frequent intervals, this factor should be borne in mind to avoid loss of money.

Fault diagnosis

Contents

1 Introduction

This Section provides an easy reference-guide to the more common faults that are likely to afflict your machine. Obviously, the opportunities are almost limitless for faults to occur as a result of obscure failures, and to try and cover all eventualities would require a book. Indeed, a number have been written on the subject.

Successful fault diagnosis is not a mysterious 'black art' but the application of a bit of knowledge combined with a systematic and logical approach to the problem. Approach any fault diagnosis by first accurately identifying the symptom and then checking through the list of possible causes, starting with the simplest or most obvious and progressing in stages to the most complex. Take nothing for granted, but above all apply liberal quantities of common sense.

The main symptom of a fault is given in the text as a major heading below which are listed, as Section headings, the various systems or areas which may contain the fault. Details of each possible cause for a fault and the remedial action to be taken are given, in brief, in the paragraphs below each Section heading. Further information should be sought in the relevant Chapter.

Engine does not start when turned over

2 No fuel flow to carburettor

- No fuel or insufficient fuel in tank.
- Fuel chamber requires priming after running dry (vacuum taps only).
- Tank filler cap air vent obstructed. Usually caused by dirt or water. Clean the vent orifice.
- Fuel tap or filter blocked. Blockage may be due to accumulation of rust or paint flakes from the tank's inner surface or of foreign matter from contaminated fuel. Remove the tap and clean it and the filter. Look also for water droplets in the fuel.
- Fuel line blocked. Blockage of the fuel line is more likely to result from a kink in the line rather than the accumulation of debris.

3 Fuel not reaching cylinder

- Float chamber not filling. Caused by float needle or float sticking in up position. This may occur after the machine has been left standing for an extended length of time allowing the fuel to evaporate. When this occurs a gummy residue is often left which hardens to a varnish-like substance. This condition may be worsened by corrosion and crystaline deposits produced prior to the total evaporation of contaminated fuel. Sticking of the float needle may also be caused by wear. In any case removal of the float chamber will be necessary for inspection and cleaning.
- Blockage in starting circuit, slow running circuit or jets. Blockage of these items may be attributable to debris from the fuel tank by-passing the filter system or to gumming up as described in paragraph 1. Water droplets in the fuel will also block jets and passages. The carburettor should be dismantled for cleaning.
- Fuel level too low. The fuel level in the float chamber is controlled by float height. The float height may increase with wear or damage but will never reduce, thus a low float height is an inherent rather than developing condition. Check the float height and make any necessary adjustment.

4 Engine flooding

- Float valve needle worn or stuck open. A piece of rust or other debris can prevent correct seating of the needle against the valve seat thereby permitting an uncontrolled flow of fuel. Similarly, a worn needle or needle seat will prevent valve closure. Dismantle the carburettor float bowl for cleaning and, if necessary, renewal of the worn components.
- Fuel level too high. The fuel level is controlled by the float height which may increase due to wear of the float needle, pivot pin or operating tang. Check the float height, and make any necessary adjustment. A leaking float will cause an increase in fuel level, and thus should be renewed.
- Cold starting mechanism. Check the choke (starter mechanism) for correct operation. If the mechanism jams in the 'On' position subsequent starting of a hot engine will be difficult.
- Blocked air filter. A badly restricted air filter will cause flooding. Check the filter and clean or renew as required. A collapsed inlet hose will have a similar effect.

5 No spark at plug

- Ignition switch not on.
- Engine stop switch off.
- Fuse blown. Check fuse for ignition circuit. See wiring diagram.
- Spark plug failure. Clean the spark plug thoroughly and reset the electrode gap. Refer to the spark plug section and the colour condition guide in Chapter 3. If the spark plug shorts internally or has sustained visible damage to the electrodes, core or ceramic insulator it should be renewed. On rare occasions a plug that appears to spark vigorously will fail to do so when refitted to the engine and subjected to the compression pressure in the cylinder.
- Spark plug cap or high tension (HT) lead faulty. Check condition and security. Replace if deterioration is evident.
- Spark plug cap loose. Check that the spark plug cap fits securely over the plug and, where fitted, the screwed terminal on the plug end is secure.
- Shorting due to moisture. Certain parts of the ignition system are susceptible to shorting when the machine is ridden or parked in wet weather. Check particularly the area from the spark plug cap back to the ignition coil. A water dispersant spray may be used to dry out waterlogged components. Recurrence of the problem can be prevented by using an ignition sealant spray after drying out and cleaning.
- Ignition or stop switch shorted. May be caused by water, corrosion or wear. Water dispersant and contact cleaning sprays may be used. If this fails to overcome the problem dismantling and visual inspection of the switches will be required.
- Shorting or open circuit in wiring. Failure in any wire connecting any of the ignition components will cause ignition malfunction. Check also that all connections are clean, dry and tight.
- Ignition coil failure. Check the coil, referring to Chapter 3.
- Defective CDI components. Check operation of the various component parts of the CDI system as described in Chapter 3. Make sure all wiring connections are sound.

6 Weak spark at plug

- Feeble sparking at the plug may be caused by any of the faults mentioned in the preceding Section other than those items in paragraphs 1 to 3. Check first the spark plug, this being the most likely culprit.

7 Compression low

- Spark plug loose. This will be self-evident on inspection, and may be accompanied by a hissing noise when the engine is turned over. Remove the plug and check that the threads in the cylinder head are not damaged. Check also that the plug sealing washer is in good condition.
- Cylinder head gasket leaking. This condition is often accompanied by a high pitched squeak from around the cylinder head and oil loss, and may be caused by insufficiently tightened cylinder head fasteners, a warped cylinder head or mechanical failure of the gasket material. Re-torqueing the fasteners to the correct specification may seal the leak in some instances but if damage has occurred this course of action will provide, at best, only a temporary cure.
- Valve not seating correctly. The failure of a valve to seat may be caused by insufficient valve clearance, pitting of the valve seat or face, carbon deposits on the valve seat or seizure of the valve stem or valve gear components. Valve spring breakage will also prevent correct valve closure. The valve clearances should be checked first and then, if these are found to be in order, further dismantling will be required to inspect the relevant components for failure.

● Cylinder, piston and ring wear. Compression pressure will be lost if any of these components are badly worn. Wear in one component is invariably accompanied by wear in another. A top end overhaul will be required.
● Piston rings sticking or broken. Sticking of the piston rings may be caused by seizure due to lack of lubrication or heating as a result of poor carburation or incorrect fuel type. Gumming of the rings may result from lack of use, or carbon deposits in the ring grooves. Broken rings result from over-revving, overheating or general wear. In either case a top-end overhaul will be required.

Engine stalls after starting

8 General causes

● Improper cold start mechanism operation. Check that the operating controls function smoothly and, where applicable, are correctly adjusted. A cold engine may not require application of an enriched mixture to start initially but may baulk without choke once firing. Likewise a hot engine may start with an enriched mixture but will stop almost immediately if the choke is inadvertently in operation.
● Ignition malfunction. See Section 9, 'Weak spark at plug'.
● Carburettor incorrectly adjusted. Maladjustment of the mixture strength or idle speed may cause the engine to stop immediately after starting. See Chapter 2.
● Fuel contamination. Check for filter blockage by debris or water which reduces, but does not completely stop, fuel flow or blockage of the slow speed circuit in the carburettor by the same agents. If water is present it can often be seen as droplets in the bottom of the float bowl. Clean the filter and, where water is in evidence, drain and flush the fuel tank and float bowl.
● Intake air leak. Check for security of the carburettor mounting and hose connections, and for cracks or splits in the hoses. Check also that the carburettor top is secure and that the vacuum gauge adaptor plug (where fitted) is tight.
● Air filter blocked or omitted. A blocked filter will cause an over-rich mixture; the omission of a filter will cause an excessively weak mixture. Both conditions will have a detrimental effect on carburation. Clean or renew the filter as necessary.
● Fuel filler cap air vent blocked. Usually caused by dirt or water. Clean the vent orifice.

Poor running at idle and low speed

9 Weak spark at plug or erratic firing

● Spark plug fouled, faulty or incorrectly adjusted. See Section 5 or refer to Chapter 3.
● Spark plug cap or high tension lead shorting. Check the condition of both these items ensuring that they are in good condition and dry and that the cap is fitted correctly.
● Spark plug type incorrect. Fit plug of correct type and heat range as given in Specifications. In certain conditions a plug of hotter or colder type may be required for normal running.
● Ignition timing incorrect. Check the ignition timing statically and dynamically, ensuring that the advance is functioning correctly.
● Faulty ignition coil. Partial failure of the coil internal insulation will diminish the performance of the coil. No repair is possible, a new component must be fitted.
● Defective CDI components. Check operation of the various component parts of the CDI system as described in Chapter 3. Make sure all wiring connections are sound.

10 Fuel/air mixture incorrect

● Intake air leak. See Section 8.
● Mixture strength incorrect. Adjust slow running mixture strength using pilot adjustment screw.
● Pilot jet or slow running circuit blocked. The carburettor should be removed and dismantled for thorough cleaning. Blow through all jets and air passages with compressed air to clear obstructions.
● Air cleaner clogged or omitted. Clean or fit air cleaner element as necessary. Check also that the element and air filter cover are correctly seated.
● Cold start mechanism in operation. Check that the choke has not been left on inadvertently and the operation is correct. Where applicable check the operating cable free play.
● Fuel level too high or too low. Check the float height and adjust as necessary. See Section 4.
● Fuel tank air vent obstructed. Obstruction usually caused by dirt or water. Clean vent orifice.
● Valve clearance incorrect. Check, and if necessary, adjust, the clearances.

11 Compression low

● See Section 7.

Acceleration poor

12 General causes

● All items as for previous Section.
● Brakes binding. Usually caused by maladjustment or partial seizure of the operating mechanism due to poor maintenance. Check brake adjustment. A bent wheel spindle or warped brake disc can produce similar symptoms.

Poor running or lack of power at high speeds

13 Weak spark at plug or erratic firing

● All items as for Section 9.
● HT lead insulation failure. Insulation failure of the HT lead and spark plug cap due to old age or damage can cause shorting when the engine is driven hard. This condition may be less noticeable, or not noticeable at all at lower engine speeds.

14 Fuel/air mixture incorrect

● All items as for Section 10, with the exception of items 2 and 3.
● Main jet blocked. Debris from contaminated fuel, or from the fuel tank, and water in the fuel can block the main jet. Clean the fuel filter, the float bowl area, and if water is present, flush and refill the fuel tank.
● Main jet is the wrong size. The standard carburettor jetting is for sea level atmospheric pressure. For high altitudes, usually above 5000 ft, a smaller main jet will be required.
● Jet needle and needle jet worn. These can be renewed individually but should be renewed as a pair. Renewal of both items requires partial dismantling of the carburettor.
● Air bleed holes blocked. Dismantle carburettor and use compressed air to blow out all air passages.
● Reduced fuel flow. A reduction in the maximum fuel flow from the fuel tank to the carburettor will cause fuel starvation, proportionate to the engine speed. Check for blockages through debris or a kinked fuel line.

15 Compression low

● See Section 7.

Knocking or pinking

16 General causes

- Carbon build-up in combustion chamber. After high mileages have been covered large accumulations of carbon may occur. These may glow red hot and cause premature ignition of the fuel/air mixture, in advance of normal firing by the spark plug. Cylinder head removal will be required to allow inspection and cleaning.
- Fuel incorrect. A low grade fuel, or one of poor quality may result in compression induced detonation of the fuel resulting in knocking and pinking noises. Old fuel can cause similar problems. A too highly leaded fuel will reduce detonation but will accelerate deposit formation in the combustion chamber and may lead to early pre-ignition as described in item 1.
- Spark plug heat range incorrect. Uncontrolled pre-ignition can result from the use of a spark plug the heat range of which is too hot.
- Weak mixture. Overheating of the engine due to a weak mixture can result in pre-ignition occurring where it would not occur when engine temperature was within normal limits. Maladjustment, blocked jets or passages and air leaks can cause this condition.

Overheating

17 Firing incorrect

- Spark plug fouled, defective or maladjusted. See Section 5.
- Spark plug type incorrect. Refer to the Specifications and ensure that the correct plug type is fitted.
- Incorrect ignition timing. Timing that is far too much advanced or far too much retarded will cause overheating. Check the ignition timing is correct.
- Defective CDI components. Check operation of the various component parts of the CDI system as described in Chapter 3. Make sure all wiring connections are sound.

18 Fuel/air mixture incorrect

- Slow speed mixture strength incorrect. Adjust pilot air screw.
- Main jet wrong size. The carburettor is jetted for sea level atmospheric conditions. For high altitudes, usually above 5000 ft, a smaller main jet will be required.
- Air filter badly fitted or omitted. Check that the filter element is in place and that it and the air filter box cover are sealing correctly. Any leaks will cause a weak mixture.
- Induction air leaks. Check the security of the carburettor mountings and hose connections, and for cracks and splits in the hoses. Check also that the carburettor top is secure and that the vacuum gauge adaptor plug (where fitted) is tight.
- Fuel level too low. See Section 3.
- Fuel tank filler cap air vent obstructed. Clear blockage.

19 Lubrication inadequate

- Engine oil too low. Not only does the oil serve as a lubricant by preventing friction between moving components, but it also acts as a coolant. Check the oil level and replenish.
- Engine oil overworked. The lubricating properties of oil are lost slowly during use as a result of changes resulting from heat and also contamination. Always change the oil at the recommended interval.
- Engine oil of incorrect viscosity or poor quality. Always use the recommended viscosity and type of oil.

20 Miscellaneous causes

- Engine fins clogged. A build-up of mud in the cylinder head and cylinder barrel cooling fins will decrease the cooling capabilities of the fins. Clean the fins as required.

Clutch operating problems

21 Clutch slip

- Friction plates worn or warped. Overhaul clutch assembly, replacing plates out of specification (Chapter 1).
- Steel plates worn or warped. Overhaul clutch assembly, replacing plates out of specification (Chapter 1).
- Clutch springs broken or worn. Old or heat-damaged (from slipping clutch) springs should be replaced with new ones (Chapter 1).
- Clutch release not adjusted properly. See Routine Maintenance.
- Clutch release mechanism defective. Worn or damaged parts in the clutch release mechanism could include the shaft, cam, actuating arm or pivot. Replace parts as necessary (Chapter 1).
- Clutch hub and outer drum worn. Severe indentation by the clutch plate tangs of the channels in the hub and drum will cause snagging of the plates preventing correct engagement. If this damage occurs, renewal of the worn components is required.
- Lubricant incorrect. Use of a transmission lubricant other than that specified may allow the plates to slip.

22 Clutch drag

- Clutch plates warped or damaged. This will cause a drag on the clutch, causing the machine to creep. Overhaul clutch assembly (Chapter 1).
- Clutch spring tension uneven. Usually caused by a sagged or broken spring. Check and replace springs (Chapter 1).
- Engine oil deteriorated. Badly contaminated engine oil and a heavy deposit of oil sludge and carbon on the plates will cause plate sticking. The oil recommended for this machine is of the detergent type, therefore it is unlikely that this problem will arise unless regular oil changes are neglected.
- Engine oil viscosity too high. Drag in the plates will result from the use of an oil with too high a viscosity. In very cold weather clutch drag may occur until the engine has reached operating temperature.
- Clutch hub and outer drum worn. Indentation by the clutch plate tangs of the channels in the hub and drum will prevent easy plate disengagement. If the damage is light the affected areas may be dressed with a fine file. More pronounced damage will necessitate renewal of the components.
- Clutch housing seized to shaft. Lack of lubrication, severe wear or damage can cause the housing to seize to the shaft. Overhaul of the clutch, and perhaps the transmission, may be necessary to repair damage (Chapter 1).
- Clutch release mechanism defective. Worn or damaged release mechanism parts can stick and fail to provide leverage. Overhaul clutch cover components (Chapter 1).
- Loose clutch hub nut. Causes drum and hub misalignment, putting a drag on the engine. Engagement adjustment continually varies. Overhaul clutch assembly (Chapter 1).

Gear selection problems

23 Gear lever does not return

- Weak or broken centraliser spring. Renew the spring.
- Gearchange shaft bent or seized. Distortion of the gearchange shaft often occurs if the machine is dropped heavily on the gear lever. Provided that damage is not severe straightening of the shaft is permissible.

24 Gear selection difficult or impossible

- Clutch not disengaging fully. See Section 22.
- Gearchange shaft bent. This often occurs if the machine is dropped heavily on the gear lever. Straightening of the shaft is permissible if the damage is not too great.
- Gearchange arms, pawls or pins worn or damaged. Wear or

breakage of any of these items may cause difficulty in selecting one or more gears. Overhaul the selector mechanism.

- Gearchange shaft centraliser spring maladjusted. This is often characterised by difficulties in changing up or down, but rarely in both directions. Adjust the centraliser anchor bolt as described in Chapter 1.
- Gearchange arm spring broken. Renew spring.
- Gearchange drum stopper cam or detent plunger damage. Failure, rather than wear, of these items may jam the drum thereby preventing gearchanging. The damaged items must be renewed.
- Selector forks bent or seized. This can be caused by dropping the machine heavily on the gearchange lever or as a result of lack of lubrication. Though rare, bending of a shaft can result from a missed gearchange or false selection at high speed.
- Selector fork end and pin wear. Pronounced wear of these items and the grooves in the gearchange drum can lead to imprecise selection and, eventually, no selection. Renewal of the worn components will be required.
- Structural failure. Failure of any one component of the selector rod and change mechanism will result in improper or fouled gear selection.

25 Jumping out of gear

- Detent plunger assembly worn or damaged. Wear of the plunger and the cam with which it locates and breakage of the detent spring can cause imprecise gear selection resulting in jumping out of gear. Renew the damaged components.
- Gear pinion dogs worn or damaged. Rounding off of the dog edges and the mating recesses in adjacent pinion can lead to jumping out of gear when under load. The gears should be inspected and renewed. Attempting to reprofile the dogs is not recommended.
- Selector forks, gearchange drum and pinion grooves worn. Extreme wear of these interconnected items can occur after high mileages especially when lubrication has been neglected. The worn components must be renewed.
- Gear pinions, bushes and shafts worn. Renew the worn components.
- Bent gearchange shaft. Often caused by dropping the machine on the gear lever.
- Gear pinion tooth broken. Chipped teeth are unlikely to cause jumping out of gear once the gear has been selected fully; a tooth which is completely broken off, however, may cause problems in this respect and in any event will cause transmission noise.

26 Overselection

- Pawl spring weak or broken. Renew the spring.
- Detent plunger worn or broken. Renew the damaged items.
- Stopper arm spring worn or broken. Renew the spring.
- Gearchange arm stop pads worn. Repairs can be made by welding and reprofiling with a file.
- Selector limiter claw components (where fitted) worn or damaged. Renew the damaged items.

Abnormal engine noise.

27 Knocking or pinking

- See Section 16.

28 Piston slap or rattling from cylinder

- Cylinder bore/piston clearance excessive. Resulting from wear, partial seizure or improper boring during overhaul. This condition can often be heard as a high, rapid tapping noise when the engine is under little or no load, particularly when power is just beginning to be applied. Reboring to the next correct oversize should be carried out and a new oversize piston fitted.
- Connecting rod bent. This can be caused by over-revving, trying to start a very badly flooded engine (resulting in a hydraulic lock in the cylinder) or by earlier mechanical failure such as a dropped valve. Attempts at straightening a bent connecting rod are not recommended. Careful inspection of the crankshaft should be made before renewing the damaged connecting rod.
- Gudgeon pin, piston boss bore or small-end bearing wear or seizure. Excess clearance or partial seizure between normal moving parts of these items can cause continuous or intermittent tapping noises. Rapid wear or seizure is caused by lubrication starvation resulting from an insufficient engine oil level or oilway blockage.
- Piston rings worn, broken or sticking. Renew the rings after careful inspection of the piston and bore.

29 Valve noise or tapping from the cylinder head

- Valve clearance incorrect. Adjust the clearances with the engine cold.
- Valve spring broken or weak. Renew the spring set.
- Camshaft or cylinder head worn or damaged. The camshaft lobes are the most highly stressed of all components in the engine and are subject to high wear if lubrication becomes inadequate. The bearing surfaces on the camshaft and cylinder head are also sensitive to a lack of lubrication. Lubrication failure due to blocked oilways can occur, but over-enthusiastic revving before engine warm-up is complete is the usual cause.
- Rocker arm or spindle wear. Rapid wear of a rocker arm, and the resulting need for frequent valve clearance adjustment, indicates breakthrough or failure of the surface hardening on the rocker arm tips. Similar wear in the cam lobes can be expected. Renew the worn components after checking for lubrication failure.
- Worn camshaft drive components. A rustling noise or light tapping which is not improved by correct re-adjustment of the cam chain tension can be emitted by a worn cam chain or worn sprockets and chain. If uncorrected, subsequent cam chain breakage may cause extensive damage. The worn components must be renewed before wear becomes too far advanced.

30 Other noises

- Big-end bearing wear. A pronounced knock from within the crankcase which worsens rapidly is indicative of big-end bearing failure as a result of extreme normal wear or lubrication failure. Remedial action in the form of a bottom end overhaul should be taken; continuing to run the engine will lead to further damage including the possibility of connecting rod breakage.
- Main bearing failure. Extreme normal wear or failure of the main bearings is characteristically accompanied by a rumble from the crankcase and vibration felt through the frame and footrests. Renew the worn bearings and carry out a very careful examination of the crankshaft.
- Crankshaft excessively out of true. A bent crank may result from over-revving or damage from an upper cylinder component or gearbox failure. Damage can also result from dropping the machine on either crankshaft end. Straightening of the crankshaft is not possible in normal circumstances; a replacement item should be fitted.
- Engine mounting loose. Tighten all the engine mounting nuts and bolts.
- Cylinder head gasket leaking. The noise most often associated with a leaking head gasket is a high pitched squeaking, although any other noise consistent with gas being forced out under pressure from a small orifice can also be emitted. Gasket leakage is often accompanied by oil seepage from around the mating joint or from the cylinder head holding down bolts and nuts. Leakage into the cam chain tunnel or oil return passages will increase crankcase pressure and may cause oil leakage at joints and oil seals. Also, oil contamination will be accelerated. Leakage results from insufficient or uneven tightening of the cylinder head fasteners, or from random mechanical failure. Retightening to the correct torque figure will, at best, only provide a temporary cure. The gasket should be renewed at the earliest opportunity.
- Exhaust system leakage. Popping or crackling in the exhaust

system, particularly when it occurs with the engine on the overrun, indicates a poor joint either at the cylinder port or at the exhaust pipe/silencer connection. Failure of the gasket or looseness of the clamp should be looked for.

Abnormal transmission noise

31 Clutch noise

- Clutch outer drum/friction plate tang clearance excessive.
- Clutch outer drum/spacer clearance excessive.
- Clutch outer drum/thrust washer clearance excessive.
- Primary drive gear teeth worn or damaged.
- Clutch shock absorber assembly worn or damaged.

32 Transmission noise

- Bearing or bushes worn or damaged. Renew the affected components.
- Gear pinions worn or chipped. Renew the gear pinions.
- Metal chips jammed in gear teeth.This can occur when pieces of metal from any failed component are picked up by a meshing pinion. The condition will lead to rapid bearing wear or early gear failure.
- Engine/transmission oil level too low. Top up immediately to prevent damage to gearbox and engine.
- Gearchange mechanism worn or damaged. Wear or failure of certain items in the selection and change components can induce mis-selection of gears (see Section 24) where incipient engagement of more than one gear set is promoted. Remedial action, by the overhaul of the gearbox, should be taken without delay.
- Loose gearbox chain sprocket. Remove the sprocket and check for impact damage to the splines of the sprocket and shaft. Excessive slack between the splines will promote loosening of the securing nut; renewal of the worn components is required. When retightening the nut ensure that it is tightened fully and that, where fitted, the lock washer is bent up against one flat of the nut.
- Chain snagging on cases or cycle parts. A badly worn chain or one that is excessively loose may snag or smack against adjacent components.

Exhaust smokes excessively

33 White/blue smoke (caused by oil burning)

- Piston rings worn or broken. Breakage or wear of any ring, but particularly the oil control ring, will allow engine oil past the piston into the combustion chamber. Overhaul the cylinder barrel and piston.
- Cylinder cracked, worn or scored. These conditions may be caused by overheating, lack of lubrication, component failure or advanced normal wear. The cylinder barrel should be renewed or rebored and the next oversize piston fitted.
- Valve oil seal damaged or worn. This can occur as a result of valve guide failure or old age. The emission of smoke is likely to occur when the throttle is closed rapidly after acceleration, for instance, when changing gear. Renew the valve oil seals and, if necessary, the valve guides.
- Valve guides worn. See the preceding paragraph.
- Engine oil level too high. This increases the crankcase pressure and allows oil to be forced past the piston rings. Often accompanied by seepage of oil at joints and oil seals.
- Cylinder head gasket blown between cam chain tunnel or oil return passage. Renew the cylinder head gasket.
- Abnormal crankcase pressure. This may be caused by blocked breather passages or hoses causing back-pressure at high engine revolutions.

34 Black smoke (caused by over-rich mixture)

- Air filter element clogged. Clean or renew the element.
- Main jet loose or too large. Remove the float chamber to check for tightness of the jet. If the machine is used at high altitudes rejetting will be required to compensate for the lower atmospheric pressure.
- Cold start mechanism jammed on. Check that the mechanism works smoothly and correctly and that, where fitted, the operating cable is lubricated and not snagged.
- Fuel level too high. The fuel level is controlled by the float height which can increase as a result of wear or damage. Remove the float bowl and check the float height. Check also that floats have not punctured; a punctured float will loose buoyancy and allow an increased fuel level.
- Float valve needle stuck open. Caused by dirt or a worn valve. Clean the float chamber or renew the needle and, if necessary, the valve seat.

Poor handling or roadholding

35 Directional instability

- Steering head bearing adjustment too tight. This will cause rolling or weaving at low speeds. Re-adjust the bearings.
- Steering head bearing worn or damaged. Correct adjustment of the bearing will prove impossible to achieve if wear or damage has occurred. Inconsistent handling will occur including rolling or weaving at low speed and poor directional control at indeterminate higher speeds. The steering head bearing should be dismantled for inspection and renewed if required. Lubrication should also be carried out.
- Bearing races pitted or dented. Impact damage caused, perhaps, by an accident or riding over a pot-hole can cause indentation of the bearing, usually in one position. This should be noted as notchiness when the handlebars are turned. Renew and lubricate the bearings.
- Steering stem bent. This will occur only if the machine is subjected to a high impact such as hitting a curb or a pot-hole. The lower yoke/stem should be renewed; do not attempt to straighten the stem.
- Front or rear tyre pressures too low.
- Front or rear tyre worn. General instability, high speed wobbles and skipping over white lines indicates that tyre renewal may be required. Tyre induced problems, in some machine/tyre combinations, can occur even when the tyre in question is by no means fully worn.
- Swinging arm bearings worn. Difficulty in holding line, particularly when cornering or when changing power settings indicates wear in the swinging arm bearings. The swinging arm should be removed from the machine and the bearings renewed.
- Swinging arm flexing. The symptoms given in the preceding paragraph will also occur if the swinging arm fork flexes badly. This can be caused by structural weakness as a result of corrosion, fatigue or impact damage, or because the rear wheel spindle is slack.
- Wheel bearings worn. Renew the worn bearings.
- Loose wheel spokes. The spokes should be tightened evenly to maintain tension and trueness of the rim.
- Tyres unsuitable for machine. Not all available tyres will suit the characteristics of the frame and suspension, indeed, some tyres or tyre combinations may cause a transformation in the handling characteristics. If handling problems occur immediately after changing to a new tyre type or make, revert to the original tyres to see whether an improvement can be noted. In some instances a change to what are, in fact, suitable tyres may give rise to handling deficiences. In this case a thorough check should be made of all frame and suspension items which affect stability.
- Front suspension units damaged or leaking. Remove from forks and examine closely (see Chapter4). Renew if suspect.
- Wear or stiffness in front suspension link pivots. Dismantle, clean and lubricate or renew pivots as necessary. See Chapter 4 for details.

36 Steering bias to left or right

- Rear wheel out of alignment. Caused by uneven adjustment of chain tensioner adjusters allowing the wheel to be askew in the fork ends. A bent rear wheel spindle will also misalign the wheel in the swinging arm.
- Wheels out of alignment. This can be caused by impact damage to the frame, swinging arm, wheel spindles or front forks. Although occasionally a result of material failure or corrosion it is usually as a result of a crash.

37 Handlebar vibrates or oscillates

- Tyres worn or out of balance. Either condition, particularly in the front tyre, will promote shaking of the fork assembly and thus the handlebars. A sudden onset of shaking can result if a balance weight is displaced during use.
- Tyres badly positioned on the wheel rims. A moulded line on each wall of a tyre is provided to allow visual verification that the tyre is correctly positioned on the rim. A check can be made by rotating the tyre; any misalignment will be immediately obvious.
- Wheel rims warped or damaged. Inspect the wheels for runout as described in Chapter 5.
- Swinging arm bushes worn. Renew the bushes.
- Wheel bearings worn. Renew the bearings.
- Steering head bearings incorrectly adjusted. Vibration is more likely to result from bearings which are too loose rather than too tight. Re-adjust the bearings.
- Loose fork component fasteners. Loose nuts and bolts holding the fork legs, wheel spindle, mudguards or steering stem can promote shaking at the handlebars. Fasteners on running gear such as the forks and suspension should be check tightened occasionally to prevent dangerous looseness of components occurring.
- Engine mounting bolts loose. Tighten all fasteners.

38 Poor rear suspension performance

- Rear suspension unit damper worn out or leaking. The damping performance of most rear suspension units falls off with age. This is a gradual process, and thus may not be immediately obvious. Indications of poor damping include hopping of the rear end when cornering or braking, and a general loss of positive stability. See Chapter 4.
- Weak rear springs. If the suspension unit springs fatigue they will promote excessive pitching of the machine and reduce the ground clearance when cornering. Although replacement springs are available separately from the rear suspension damper unit it is probable that if spring fatigue has occurred the damper units will also require renewal.
- Swinging arm flexing or bushes worn. See Sections 35 and 36.
- Bent suspension unit damper rod. This is likely to occur only if the machine is dropped or if seizure of the piston occurs. If either happens the suspension units should be renewed as a pair.

Abnormal frame and suspension noise

39 Front end noise

- Spring weak or broken. Makes a clicking or scraping sound.
- Steering head bearings loose or damaged. Clicks when braking. Check, adjust or replace (Chapter 4).

40 Rear suspension noise

- Fluid level too low. Leakage of a suspension unit, usually evident by oil on the outer surfaces, can cause a spurting noise. The suspension units should be renewed as a pair.
- Defective rear suspension unit with internal damage. Renew the suspension units as a pair.

Brake problems

41 Brakes are spongy or ineffective

- Brake cable deterioration. Damage to the outer cable by stretching or being trapped will give a spongy feel to the brake lever. The cable should be renewed. A cable which has become corroded due to old age or neglect of lubrication will partially seize making operation very heavy. Lubrication at this stage may overcome the problem but the fitting of a new cable is recommended.
- Worn brake linings. Determine lining wear using the external brake wear indicator on the brake backplate, or by removing the wheel and withdrawing the brake backplate. Renew the shoe/lining units as a pair if the linings are worn below the recommended limit.
- Worn brake camshaft. Wear between the camshaft and the bearing surface will reduce brake feel and reduce operating efficiency. Renewal of one or both items will be required to rectify the fault.
- Worn brake cam and shoe ends. Renew the worn components.
- Linings contaminated with dust or grease. Any accumulations of dust should be cleaned from the brake assembly and drum using a petrol dampened cloth. Do not blow or brush off the dust because it is asbestos based and thus harmful if inhaled. Light contamination from grease can be removed from the surface of the brake linings using a solvent; attempts at removing heavier contamination are less likely to be successful because some of the lubricant will have been absorbed by the lining material which will severely reduce the braking performance.

42 Brake drag

- Incorrect adjustment. Re-adjust the brake operating mechanism.
- Drum warped or oval. This can result from overheating, impact or uneven tension of the wheel spokes. The condition is difficult to correct, although if slight ovality only occurs, skimming the surface of the brake drum can provide a cure. This is work for a specialist engineer. Renewal of the complete wheel hub is normally the only satisfactory solution.
- Weak brake shoe return springs. This will prevent the brake lining/shoe units from pulling away from the drum surface once the brake is released. The springs should be renewed.
- Brake camshaft, lever pivot or cable poorly lubricated. Failure to attend to regular lubrication of these areas will increase operating resistance which, when compounded, may cause tardy operation and poor release movement.

43 Brake lever or pedal pulsates in operation

- Drums warped or oval. This can result from overheating, impact or uneven spoke tension. This condition is difficult to correct, although if slight ovality only occurs skimming the surface of the drum can provide a cure. This is work for a specialist engineer. Renewal of the hub is normally the only satisfactory solution.

44 Drum brake noise

- Drum warped or oval. This can cause intermittent rubbing of the brake linings against the drum. See the preceding Section.
- Brake linings glazed. This condition, usually accompanied by heavy lining dust contamination, often induces brake squeal. The surface of the linings may be roughened using glass-paper or a fine file.

45 Brake induced fork judder

- Worn front fork stanchions and legs, or worn or badly adjusted steering head bearings. These conditions, combined with uneven or pulsating braking as described in Section 43 will induce more or less judder when the brakes are applied, dependent on the degree of wear and poor brake operation. Attention should be given to both areas of malfunction. See the relevant Sections.

Electrical problems

46 Battery dead or weak

- Battery faulty. Battery life should not be expected to exceed 3 to 4 years, particularly where a starter motor is used regularly. Gradual sulphation of the plates and sediment deposits will reduce the battery performance. Plate and insulator damage can often occur as a result of vibration. Complete power failure, or intermittent failure, may be due to a broken battery terminal. Lack of electrolyte will prevent the battery maintaining charge.

- Battery leads making poor contact. Remove the battery leads and clean them and the terminals, removing all traces of corrosion and tarnish. Reconnect the leads and apply a coating of petroleum jelly to the terminals.
- Load excessive. If additional items such as spot lamps, are fitted, which increase the total electrical load above the maximum alternator output, the battery will fail to maintain full charge. Reduce the electrical load to suit the electrical capacity.
- Regulator/rectifier failure.
- Alternator generating coils open-circuit or shorted.
- Charging circuit shorting or open circuit. This may be caused by frayed or broken wiring, dirty connectors or a faulty ignition switch. The system should be tested in a logical manner. See Section 49.

47 Battery overcharged

- Rectifier/regulator faulty. Overcharging is indicated if the battery becomes hot or it is noticed that the electrolyte level falls repeatedly between checks. In extreme cases the battery will boil causing corrosive gases and electrolyte to be emitted through the vent pipes.
- Battery wrongly matched to the electrical circuit. Ensure that the specified battery is fitted to the machine.

48 Total electrical failure

- Fuse blown. Check the main fuse. If a fault has occurred, it must be rectified before a new fuse is fitted.
- Battery faulty. See Section 46.
- Earth failure. Check that the frame main earth strap from the battery is securely affixed to the frame and is making a good contact.
- Ignition switch or power circuit failure. Check for current flow through the battery positive lead (red) to the ignition switch. Check the ignition switch for continuity.

49 Circuit failure

- Cable failure. Refer to the machine's wiring diagram and check the circuit for continuity. Open circuits are a result of loose or corroded connections, either at terminals or in-line connectors, or because of broken wires. Occasionally, the core of a wire will break without there being any apparent damage to the outer plastic cover.
- Switch failure. All switches may be checked for continuity in each switch position, after referring to the switch position boxes incorporated in the wiring diagram for the machine. Switch failure may be a result of mechanical breakage, corrosion or water.
- Fuse blown. Refer to the wiring diagram to check whether or not a circuit fuse is fitted. Replace the fuse, if blown, only after the fault has been identified and rectified.

50 Bulbs blowing repeatedly

- Vibration failure. This is often an inherent fault related to the natural vibration characteristics of the engine and frame and is, thus, difficult to resolve. Modifications of the lamp mounting, to change the damping characteristics may help.
- Intermittent earth. Repeated failure of one bulb, particularly where the bulb is fed directly from the generator, indicates that a poor earth exists somewhere in the circuit. Check that a good contact is available at each earthing point in the circuit.

Routine maintenance

Specifications

Engine/transmission

Valve clearances (cold):	
Inlet	0.050 – 0.100 mm (0.0020 – 0.0040 in)
Exhaust	0.075 – 0.125 mm (0.0030 – 0.0050 in)
Idle speed	1700 ± 100 rpm
Spark plug gap	0.6 – 0.7 mm (0.024 – 0.028 in)

Cycle parts

Front brake lever free play	5.0 – 8.0 mm (0.2 – 0.3 in)	
Rear brake pedal free play	20 – 30 mm (0.8 – 1.2 in)	
Tyre pressures cold:	**Front**	**Rear**
Up to 75 kg/165 lb load	22 psi (1.54 kg/cm²)	28 psi (2.0 kg/cm²)
Above 75 kg/165 lb load – early (35T) T80 models	22 psi (1.54 kg/cm²)	32 psi (2.25 kg/cm²)
Above 75 kg/165 lb load – all T50 and later (2FL) T80 models	22 psi (1.54 kg/cm²)	40 psi (2.8 kg/cm²)

Recommended lubricants

Engine/transmission unit:	
Capacity – dry	1.0 litre (1.8 Imp pint)
Capacity – at oil change	0.85 litre (1.5 Imp pint)
Oil grade	SAE 10W/40, type SE or SF motor oil
Control cables	General purpose lubricating oil
Stand and lever pivots	General purpose grease
Speedometer drive	General purpose grease
Wheel bearings	High melting-point grease
Brake cams	High melting-point grease
Final drive chain	Aerosol chain grease

Introduction

Periodic routine maintenance is a continuous process which should commence immediately the machine is used. The object is to maintain all adjustments and to diagnose and rectify minor defects before they develop into more extensive, and often more expensive, problems.

It follows that if the machine is maintained properly, it will both run and perform with optimum efficiency, and be less prone to unexpected breakdowns. Regular inspection of the machine will show up any parts which are wearing, and with a little experience, it is possible to obtain the maximum life from any one component, renewing it when it becomes so worn that it is liable to fail.

Regular cleaning can be considered as important as mechanical maintenance. This will ensure that all the cycle parts are inspected regularly and are kept free from accumulations of road dirt and grime.

Cleaning is especially important during the winter months, despite its appearance of being a thankless task which very soon seems pointless. On the contrary, it is during these months that the paintwork, chromium plating, and the alloy casings suffer the ravages of abrasive grit, rain and road salt. A couple of hours spent weekly on cleaning the machine will maintain its appearance and value, and highlight small points, like chipped paint, before they become a serious problem.

The various maintenance tasks are described under their respective mileage and calendar headings, and are accompanied by diagrams and photographs where pertinent.

It should be noted that the intervals between each maintenance task serve only as a guide. As the machine gets older, or if it is used under particularly arduous conditions, it is advisable to reduce the period between each check.

For ease of reference, most service operations are described in detail under the relevant heading. However, if further general information is required, this can be found under the pertinent Section heading and Chapter in the main text.

Although no special tools are required for routine maintenance, a good selection of general workshop tools is essential. Included in the tools must be a range of metric ring or combination spanners, a selection of crosshead screwdrivers, and two pairs of circlip pliers, one external opening and the other internal opening. Additionally, owing to the extreme tightness of most casing screws on Japanese machines, an impact screwdriver, together with a choice of large or small cross-head screw bits, is absolutely indispensable. This is particularly so if the engine has not been dismantled since leaving the factory.

Note: The service intervals given in the following sections are as specified in Yamaha's official literature, and will be as found in the machine's handbook. It should be noted, however, that the UK importers specify slightly different mileage/time recommendations, particularly for the period that the machine is under warranty. These are as shown below:

3 monthly or every 4000 miles (6000 km)
6 monthly or every 8000 miles (12 000 km)

Removing the legshields

Access to a number of the engine ancillaries and other parts is simplified if the moulded plastic legshield assembly is removed. This is not a lengthy process, and may be considered worthwhile even where access holes are provided, in view of the improved visibility offered. Where legshield removal is essential, this will be indicated in the text.

The legshield moulding is retained by a total of six cross head screws and washers, the upper right-hand screw doubling as a mounting point for the luggage hook. Remove the screws, washers and spacers, noting their relative positions. The legshield can now be lifted away from the frame.

Care should be taken when handling the legshield moulding; it is resilient, but is easily scratched. Place it on soft cloth in a safe place when removed, and take care not to score the surface during removal or fitting. Remember to refit any spacers during installation, and fit all fasteners finger-tight before final tightening. Check that the legshield is aligned correctly, then tighten the screws evenly. Beware of overtightening.

Pre-ride inspection

The operations shown below should be carried out prior to riding the machine each day. The procedure should take only a few minutes, and will significantly reduce the risk of unexpected failures or breakdowns whilst on the road. It should be borne in mind that the T80 models are designed to require minimal repair maintenance. Whilst this means that the more time-consuming service requirements occur at six-monthly intervals, the condition of much of the machine is reliant on the pre-ride checks, so do not be tempted to ignore the pre-ride inspection sequence shown below.

1 Checking the engine/transmission oil level

With the engine cold and the machine standing on a level surface, check the level of the engine oil via the sight glass in the engine right-hand outer cover. The oil level should be between the high and low level lines. If necessary, add SAE 10W/40 motor oil (type SE or SF) to bring the level above the minimum mark. Refit the plug, screwing it home firmly.

2 Checking the operation of the front and rear brakes

Check that the front brake lever and rear brake pedal operate smoothly and evenly, and that the control cables are correctly adjusted (see 6 monthly/4000 mile heading for details). Turn the ignition switch on and check that the brake lamp switches operate correctly.

3 Checking the operation of the suspension and steering

Check the suspension by bouncing the machine up and down. Investigate and rectify any signs of stiffness or abnormal noises (Chapter 4). Check the steering operation by turning the handlebar from lock to lock with the machine on its centre stand. If stiffness or excessive free play are noted, check for dry, damaged or badly adjusted steering head bearings as described in Chapter 4.

4 Checking throttle operation

Check that the throttle control operates smoothly and easily, and that it returns when released. The throttle cable adjuster should be set to give 5 – 8 mm (0.2 – 0.3 in) free play at the outer edge of the twistgrip flange. Further details will be found under the 6 monthly/4000 mile heading.

5 Checking the tyre pressures and condition

Check the tyre pressures and condition, looking for nails or stones in the tread or damage to the sidewalls, as well as normal tread wear. In addition to local legal requirements, if the tread is worn to the manufacturer's service limit of 1 mm or less at any point, renew the tyre. If damage is found, either renew the tyre or seek professional advice as to whether the tyre is safe for continued use. Note that when new tyres are fitted, Yamaha recommend the fitment of Inoue tyres of the approved size and do not guarantee the ride or handling characteristics of the machine on tyres from other suppliers. If in doubt, seek the assurance of the tyre supplier when choosing alternative brands.

6 Checking the battery electrolyte level

Remove the right-hand side panel and check the battery electrolyte level. If below the minimum level, top up using distilled water (see 6 monthly/4000 mile service interval for details).

7 Lighting and electrical system checks

Start the engine and allow it to warm up whilst checking that all lights operate normally. Check that the horn works and that there is sufficient fuel in the tank for the intended journey.

8 Checking fittings, fasteners and stand operation

Check around the machine for any signs of loose fittings or slack or missing securing screws, bolts or nuts. Tighten or replace as necessary. Check that the stand retracts smoothly and completely and that the stand pivots are sound and secure.

9 Checks whilst riding

When riding the machine initially, note any obvious minor faults such as stiff or erratic switch operation or loose fittings or fasteners and attend to these promptly. Try to check periodically that items like the brake, tail and turn signal lamps are working normally. If minor problems such as an occasional misfire are noted, investigate the problem as soon as possible; this can often avoid a breakdown at a later date.

Engine oil level must lie between the upper and lower level lines

Check that rear suspension unit spring preload is at the same setting on each unit

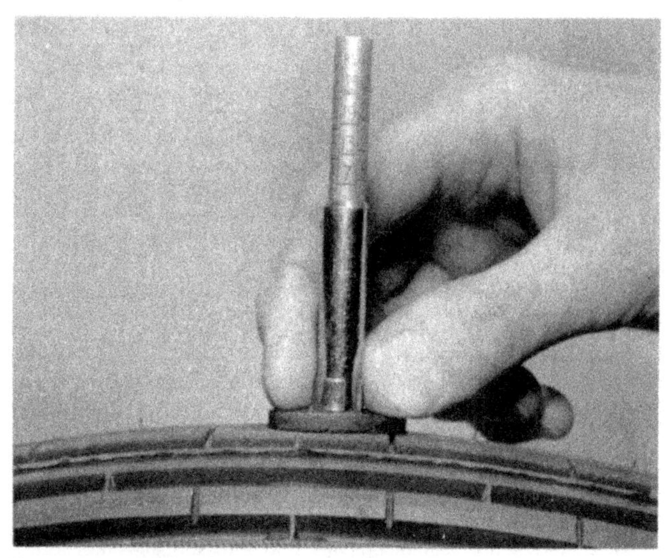

Use a tyre tread depth gauge to assess tyre wear

Unscrew the hexagon-headed inspection caps to gain access to the valve adjusters

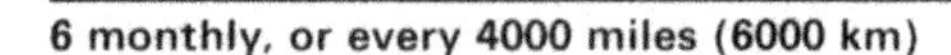

6 monthly, or every 4000 miles (6000 km)

After carrying out the normal pre-ride checks, excluding those which require the engine to be run, carry out the following operations. Follow the sequence indicated so that those operations requiring a cold engine are completed before those requiring a hot engine.

1 Checking the cam chain tensioner adjustment and valve clearances

The cam chain tension should be checked and adjusted at the same time as the valve clearances are checked. The clearance between the top of each valve stem and its rocker must be checked at the prescribed interval to ensure that it remains at the specified figure. If the gap becomes too great, the valves will not be opened fully and the engine will run less efficiently. An excessive degree of noise will be produced. Conversely, an insufficient gap will mean loss of power and eventually will cause the affected valve to burn out, necessitating a cylinder head overhaul. The clearance check must be made with the engine cold, so leave the machine to stand overnight before attempting to carry out this operation. The legshield should be removed for this check (see above).

Cam chain tensioner adjustment

Remove the gearchange pedal and then the crankcase left-hand outer cover to expose the alternator rotor. Remove the circular inspection cover on the left-hand side of the cylinder head, which is retained by two bolts. Using the alternator rotor, turn the crankshaft assembly anticlockwise until the 'T' mark on the rotor edge aligns with the corresponding fixed index mark cast into the crankcase edge. Check that the alignment mark on the camshaft sprocket aligns with its index mark; if necessary, rotate the crankshaft through a further full turn until it coincides.

Remove the tensioner lockscrew domed nut, then slacken the tensioner pushrod locknut and screw to allow the assembly free movement. Now turn the crankshaft, again anticlockwise, until the locating tang at the centre of the cam sprocket is in line with the tensioner pushrod locknut and screw. The spring in the tensioner will set the mechanism automatically to allow the correct cam chain free play. Tighten the lockscrew, securing the setting by tightening the locknut to 0.7 kgf m (5.1 lbf ft). Refit the domed nut and tighten it to 0.5 kgf m (3.6 lbf ft). Before refitting the inspection and side covers and the gearchange pedal, carry out the valve clearance adjustment procedure described below.

Valve clearance adjustment

Unscrew the two hexagon-headed inspection caps on the cylinder head. Turn the rotor anticlockwise until the 'T' mark on its edge aligns with the index mark on the adjacent crankcase edge. Check that the engine is at TDC (top dead centre) on the compression stroke. This can be established by grasping each rocker arm in turn by its hexagon head and pushing and pulling it. A small clearance should be detected on each one; if one seems tight or the valve is partially open, turn the rotor through one full turn, realign the 'T' mark and check again.

Measure the clearance between each rocker arm adjuster screw and the top of its valve stem using feeler gauges. The clearance is indicated by the gauge which should be a slight sliding fit between the two components. Note the clearance and then repeat on the remaining valve. The specified figure is as follows. If adjustment is required, proceed as described below.

Inlet	0.05 – 0.10 mm (0.002 – 0.004 in)
Exhaust	0.075 – 0.125 mm (0.003 – 0.005 in)

To carry out any adjustment, a ring spanner should be used on the locknut, whilst the adjuster can be turned with the appropriate size of magneto spanner. Slacken the locknut by about 1/2 turn, holding it in this position whilst the adjuster is turned to give the required setting. Check the clearance with feeler gauges, then holding the adjuster, secure the locknut. Re-check the clearance and make any corrections required before repeating the procedure on the remaining valve. Finally, refit the inspection covers, the crankcase left-hand cover and the gearchange pedal.

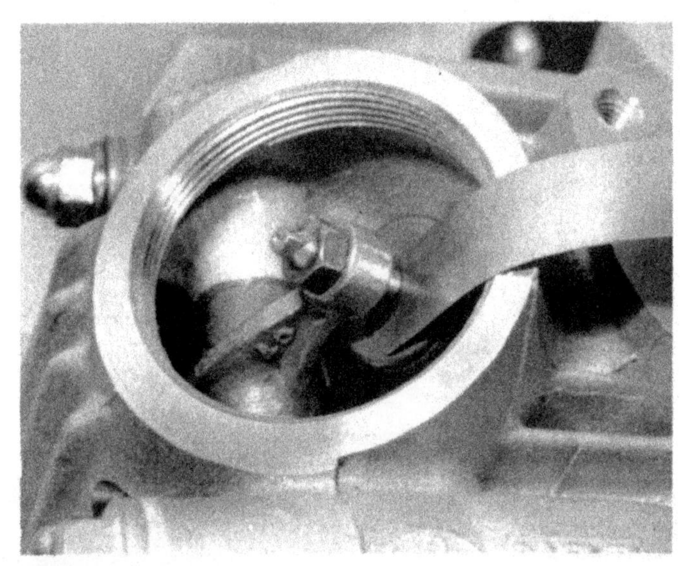

Use feeler gauges to measure the clearance between each rocker arm and valve stem

2 Checking the spark plug

The spark plug should be removed for examination. Access to the plug is possible through the cutout provided in the legshield. Pull off the plug cap and unscrew the plug from the cylinder head, using a proper plug spanner. The plug spanner supplied with the machine is quite adequate for this purpose, but if it has been lost, it should be noted that the 'standard' plug spanners and sockets supplied by most tool and accessory stores are to suit plugs with a 14 mm thread (21 mm hexagon) and will not fit; a spanner or socket to fit plugs with a 10 mm thread (16 mm hexagon) is required.

Given the low cost of a new plug, the author recommends that a new plug is fitted as a matter of course. This ensures that it is almost certain that plug failure will never be experienced. On the other hand, the used plug can be cleaned and refitted, provided that the electrodes have not worn too badly. Before attempting to clean or renew the plug; refer to the colour pages which describe plug condition in Chapter 3. This can provide a useful guide to the condition of the engine in general.

A used plug can be cleaned by abrasive blasting in a specially designed machine, either a full-sized workshop type or one of the inexpensive home versions. Note however, that the small size of the plug may make this impossible on many machines. The alternative is to carefully scrape away the carbon deposits, taking great care to avoid damage to the electrodes or the ceramic insulator. Whichever method is chosen, make sure that all traces of loose material and any abrasive dust is removed completely or engine damage will almost certainly be caused. Finally, examine the electrode surfaces, and if necessary file them flat and parallel using a small magneto file.

Whether a new or a cleaned plug is to be fitted, always set the electrode gap carefully to 0.6 – 0.7 mm (0.024 – 0.028 in). Measure the gap using feeler gauges, and if adjustment is required, bend the outer earth electrode only to achieve the required gap. On no account attempt to bend the centre electrode or the ceramic insulator will be cracked or broken off.

Check that the plug threads are clean, then wipe them with molybdenum disulphide or PBC grease prior to fitting. This ensures that the plug will fit easily, and avoids the risk of stripped plug threads in the cylinder head at a later date. Fit the plug finger-tight only, then tighten with the plug spanner by a further 1/4 – 1/2 turn, just enough to seat the plug.

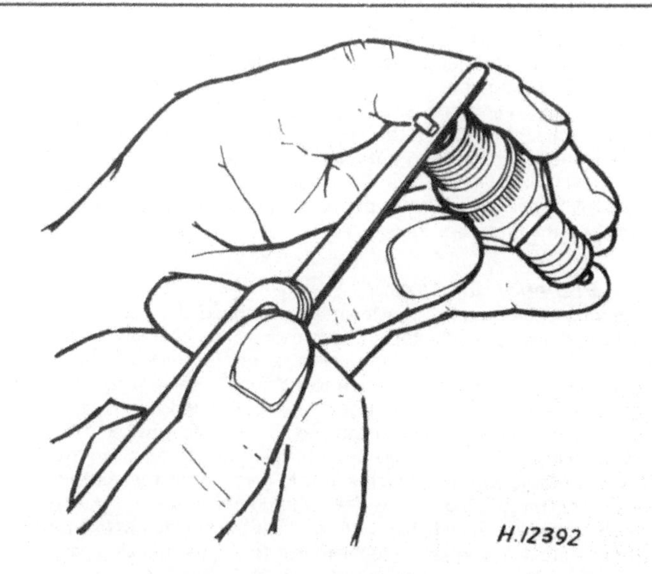

Measuring the spark plug gap

3 Cleaning the air filter element

The air filter element is housed in a moulded plastic casing mounted below the main frame tube, just to the rear of the steering head. To gain access, it is preferable to remove legshield assembly as described above, though it can be reached from the underside with the legshield in place.

The air filter cover is secured to the casing by four crosshead screws. Release the screws, then slacken the clamp which secures the connecting hose to the carburettor. (This may prove difficult if the clamp was badly positioned when it was last fitted. If so, removal of the legshield may prove essential.) Turn the steering to full lock in either direction to allow the filter cover and element to be lowered clear of the casing.

The flat foam element can be peeled away for cleaning, but check first that the foam is undamaged; a torn or holed element must be renewed. It is essential that the filter is in good condition and that the engine should never be used with the element missing. The tiny particles caught by the filter can cause rapid and expensive engine wear if they are drawn into the cylinder.

Clean the element by washing it thoroughly in petrol, taking suitable precautions to avoid any risk of fire and working outdoors, or in a well-ventilated area. Squeeze out any residual petrol and then dry the element with clean rag. Once dry, re-oil the element with SAE 10W/30 motor oil. Squeeze out the excess oil to leave the element surfaces damp but not dripping with oil. Refit the element and cover, ensuring that it seats properly.

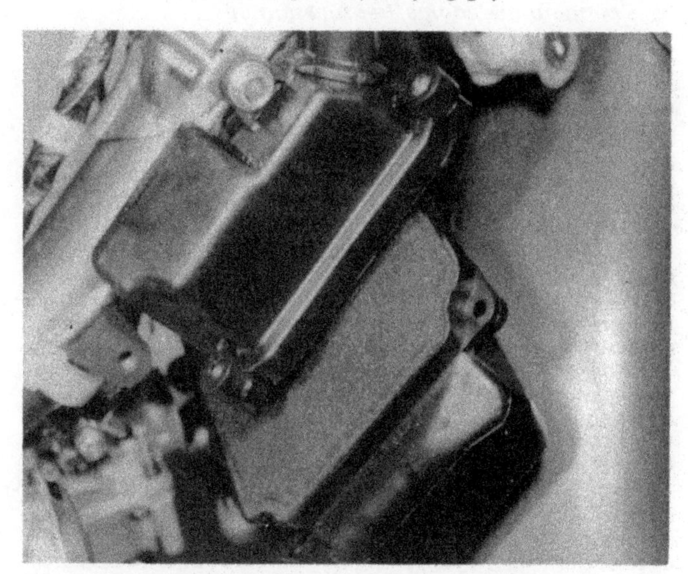
Remove legshield and separate air filter casing to gain access to foam filter element

4 Throttle cable adjustment

Check that the throttle control operates smoothly and that it closes properly when released. The throttle cable should be set so that there is 2 – 5 mm (0.08 – 0.20 in) free play measured at the twistgrip flange. To alter the adjustment, slacken the adjuster locknut and turn the adjuster in or out to obtain the correct clearance. The adjuster is located just below the twistgrip unit and is normally covered by a rubber sleeve.

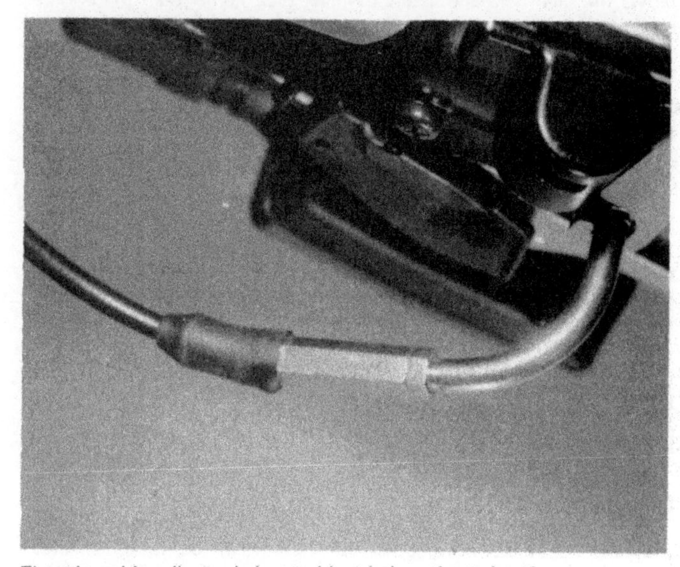
Throttle cable adjuster is located just below the twistgrip

5 Checking the fuel and vacuum pipes

Examine the fuel and vacuum pipes for signs of deterioration or splitting. If damaged near the ends they may be restored by trimming off the damaged area and refitting the pipe, but there is an obvious limit to the number of times that this can be done. If renewal is required, use only synthetic rubber tubing of the correct type. This can be obtained from most motorcycle dealers. Do not use natural rubber tubing, which is quickly attacked by fuel, or plastic tubing which is not designed to withstand the effects of fuel.

6 Carburettor adjustment

Normal carburettor adjustment is confined to checking and resetting the engine idle speed. To carry out this operation with any degree of accuracy requires the use of a test tachometer, and some owners may prefer to leave the job to a Yamaha dealer, who will have the necessary equipment at his disposal. For those possessing a suitable tachometer, the idle speed should be set to 1700 ± 100 rpm with the engine at normal operating temperature. In practice, the idle speed can be set without a tachometer to give the slowest reliable idle speed. The throttle stop screw (see photograph) can be reached via the access hole in the left-hand side of the legshield; do not disturb the pilot mixture screw which is located next to the throttle stop screw.

Set the throttle stop screw as shown – do not disturb pilot air screw (A)

7 Changing the engine/transmission oil

The engine oil should be changed at the recommended interval to ensure adequate lubrication of the engine and transmission components; failure to do so may cause premature wear or damage in these areas. Before draining the old oil, make sure that the engine is at normal operating temperature, preferably by riding the machine for several miles. This ensures that the oil drains fully and that any contaminants are suspended in the oil, rather than being left on the bottom of the crankcase.

Remove the filler plug, then place a container of at least 1 litre or 1 1/2 Imp pint below the crankcase, positioning it below the drain plug. Unscrew the drain plug and leave the oil to drain fully. While the oil is draining, remove the footrest assembly, the exhaust system and the kickstart lever. Slacken the screws which retain the crankcase right-hand outer cover and lift it away. Remove the oil strainer from its slot at the bottom of the crankcase, just forward of the clutch. Wash the strainer in petrol and check that it is in good condition before refitting it. Take suitable precautions to prevent the risk of fire when using petrol. Note that if the strainer mesh is torn or split it must be renewed. On no account use the machine with a damaged or missing strainer.

Clean the drain plug carefully, and ensure that the sealing washer is in good condition. If in any doubt, renew the washer. When the oil has stopped dripping from the drain plug hole, clean the threads with a piece of clean rag. Fit the plug finger-tight, then torque to 2.0 kgf m (14 lbf ft).

Add 0.85 litre (1.5 Imp pint) of SAE 10W/40 type SE or SF motor oil via the filler hole and screw the filler plug home. Start the engine and run it for 2 – 3 minutes, then stop it and allow it to stand for a similar time. Check the level of the oil via the sight glass, ensuring that the machine is absolutely vertical during the check. If necessary, top up to bring the oil level above the minimum mark. When the level is correct, refit the filler plug.

Top up engine/gearbox unit using SAE 10W/40 motor oil

8 Checking the clutch adjustment

The setting of the clutch pushrod adjuster should be checked at the recommended interval, or whenever clutch operation is suspect. Remove the plug which covers the adjuster screw head and slacken the locknut. Turn the adjuster screw inwards until it seats lightly, then back it off by 1/8 to 1/4 turn. Holding the screw in this position, tighten the locknut, then refit the blanking plug.

Set the clutch pushrod adjuster to 1/8 – 1/4 turns out

9 Checking the battery

Note: *Battery electrolyte contains sulphuric acid. Take care to avoid contact with eyes, skin or clothing. Refer to the 'Safety First!' section at the beginning of this manual.*

Remove the right-hand side panel to gain access to the battery. It is best to remove the battery from its recess for checking, and this is accomplished as follows. Unclip the fuseholder from the battery holder strap and disconnect the breather pipe. Remove the bolt which secures the battery earth lead and the battery tray to the frame. Unhook the battery strap to release the battery and slide it out sufficiently to allow the positive (+) lead to be released at the terminal post. Place the battery on the workbench for inspection.

The battery electrolyte level should lie between the upper and lower marks on the casing. If topping up is required, unscrew the filler caps and add only distilled (or de-mineralised) water to restore each cell to the upper level line. Refit the filler caps and clean the battery casing before refitting it in its recess.

If abnormally frequent topping up is needed, there may be a fault in the regulator/rectifier unit. See Chapter 6 for details. Repeatedly flat batteries may be due to neglect, a charging system fault, or simply that the battery has reached the end of its useful life. A well-maintained battery will normally last about three years.

10 Checking the brake shoes

The brakes are each fitted with a wear indicator which indicates the condition of the shoes without the need for dismantling for visual inspection. Apply each brake in turn and check the position of the brake arm pointer in relation to the wear limit mark on the brake backplate. If the maximum mark has been reached, the shoes are worn to the point where renewal is required. For details on removing and fitting the brake shoes, refer to Chapter 5.

11 Brake inspection and adjustment

Check the operation of both brakes. Examine the front brake cable for signs of fraying or damage to the cable outer, and renew it if suspect. Check the rear brake linkage for wear and lubricate or renew the worn parts as required.

To adjust the rear brake, set the adjuster at the wheel end to give 20 - 30 mm (0.8 - 1.2 in) free play at the pedal end. In the case of the front brake, there should be 5 - 8 mm (0.2 - 0.3 in) free play measured between the lever stock and blade. It is good practice to apply a coat of grease to the exposed adjuster threads to prevent any risk of seizure due to corrosion.

Check that the brake lamp comes on when the brake lever or pedal is operated. The rear brake switch should be adjusted so that the lamp comes on just before the brake begins to take effect. The switch is adjusted by moving the assembly up or down in relation to its mounting bracket. Hold the switch body still and turn the adjuster nuts until the desired setting is achieved.

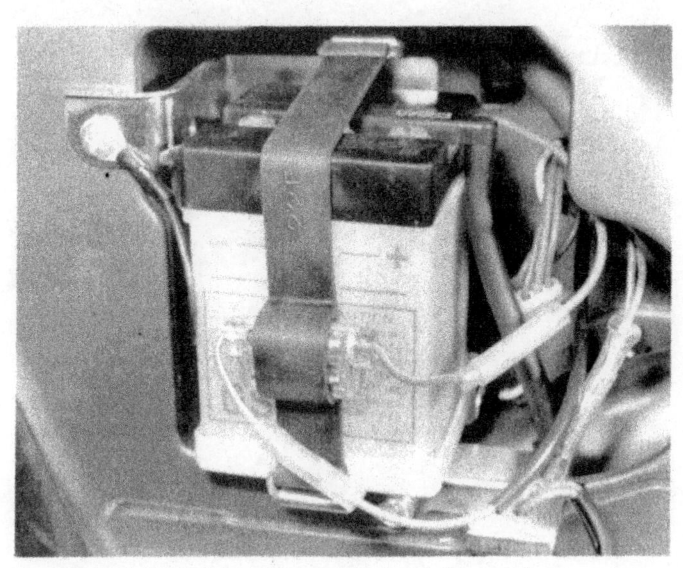

The battery is located behind the right-hand side panel

Renew shoes if pointer reaches maximum wear limit with brake applied

Set the brake cable or rod adjuster to give specified free play

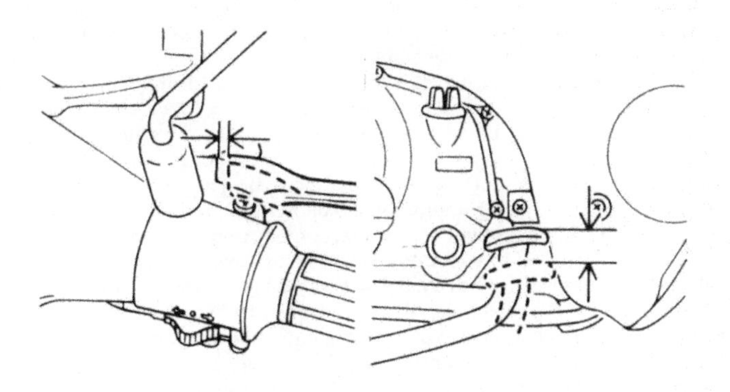

Brake adjustment settings

12 Wheel rim and spoke examination

Check both wheel rims and spokes for signs of damage or corrosion. Minor adjustments can be made at home, but badly loosened spokes or impact damage will need specialist attention. For more information, see Chapter 5.

13 Checking the suspension

Give the front and rear suspension a close visual examination, preferably after the machine has been degreased and cleaned. Check the action of the suspension, noting any obvious play or noisy operation, or any sign of oil leakage from the front and rear suspension units. Trace and rectify any fault. Check that all fasteners are tightened securely.

In the case of the front suspension, grease the suspension links via the grease nipples near the front of each one, using a general purpose grease applied with a grease gun.

With the machine on its centre stand, push and pull the swinging arm from side to side to check for free play. Any obvious free movement warrants investigation. All aspects of the suspension are covered in detail in Chapter 4.

14 Checking and adjusting the steering head bearings

Place the machine on its centre stand and ask an assistant to sit on the rear of the seat so that the front wheel is raised clear of the ground. Check that the steering moves smoothly and easily from lock to lock. If the steering feels stiff or notchy in operation, dismantle and check the steering head bearings as described in Chapter 4. Grasp the lower ends of the forks and try to push and pull them to feel for free play. Any clearance in the bearings will be magnified at the wheel end. If play is discovered, adjust the bearings as described in Chapter 4. Given that a certain amount of dismantling is necessary to carry out the adjustment, it is suggested that the bearings should be removed, cleaned and greased at the same time. It should be noted that Yamaha recommend that the bearings should be lubricated at 16 000 mile/two yearly intervals.

15 Checking and lubricating the stand pivots

Check the side (where fitted) and centre stand pivots for security and smooth operation. Lubricate the pivots, if necessary after dismantling and cleaning. Examine the stand return springs and renew them if damaged or badly rusted.

16 Changing the final drive grease

The final drive bevel gear grease must be renewed at this interval. Access involves removal of the rear wheel and bevel gearbox housing as described in Chapter 5, Section 6, and Chapter 4, Section 13.

Note that the type and quantity of grease is of particular importance to ensure the correct operation of the final drive; see Chapter 4, Section 13 for further details.

17 Checking fasteners, wiring and control cables

Give the machine a close check for loose nuts and bolts, tightening them as required to the appropriate torque setting. These are given in the specifications at the beginning of each Chapter. It is a good idea to check other items such as wiring and wiring connectors and cable condition and routing at the same time.

The various control cables should be checked for fraying or wear and renewed as required. Lubricate all cables to maintain smooth and reliable operation. Lubrication can be carried out by detaching the top of the cable and forcing oil through the cable outer. A traditional method is to construct a makeshift funnel by taping a plastic bag around the cable, filling it with oil and allowing it to drain through the cable, preferably overnight (see the accompanying line drawing).

An alternative is to use a proprietary cable oiler such as one of the hydraulic types where the tool is filled with oil and a screw tightened to force the oil along the cable. Newer versions clamp over the cable and accept the extension nozzle of most maintenance aerosol sprays. Most motorcycle dealers will stock one type or another, but if difficulty is experienced in finding a supplier, try a dealer specialising in off-road and competition machines.

Yearly, or every 8000 miles (12 000 km)

Carry out the operations listed under the previous mileage and calendar headings, then complete the following:

1 Checking the wheel bearings

The front and rear wheel bearings should be examined for signs of wear and renewed as required. The procedures are described in Chapter 5.

2 Checking the swinging arm pivot

With the machine on its centre stand check for excessive free play by attempting to push and pull the swinging arm sideways. Also check tighten the pivot shaft. Yamaha recommend that the bushes are inspected every 16 000 miles/2 years; see Chapter 4.

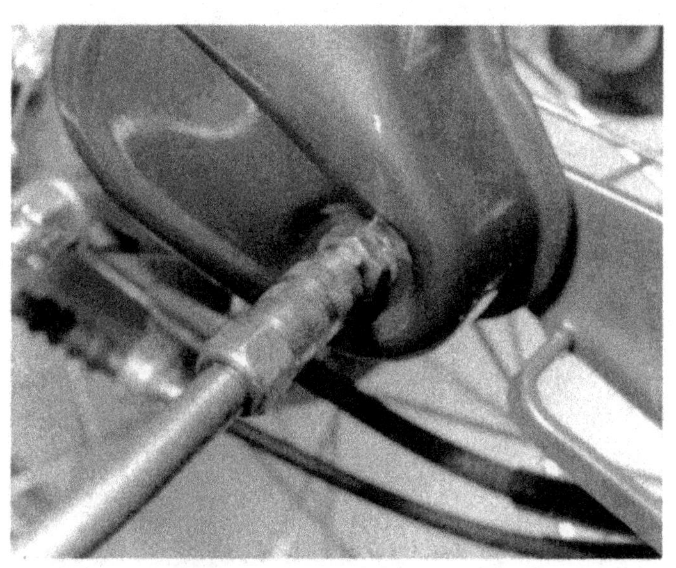

Grease the front suspension links via the grease nipples

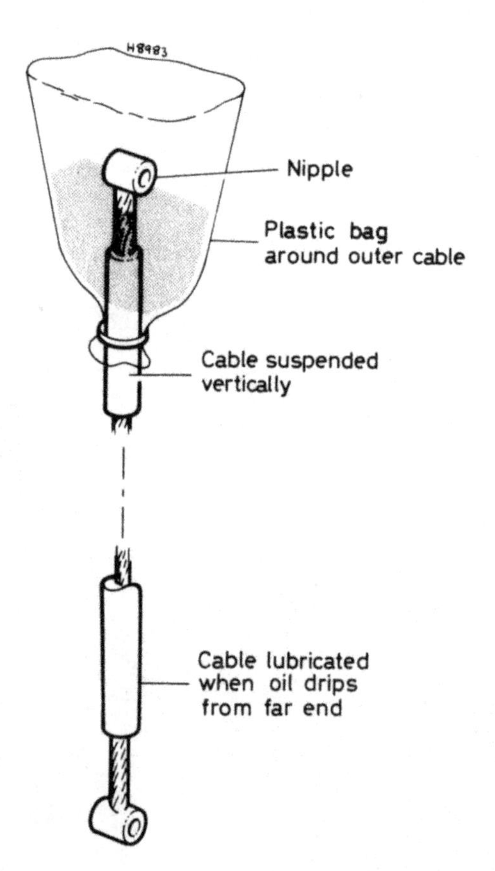

Oiling a control cable

Additional maintenance items

1 Cleaning the machine

Keeping the machine clean should be considered an important part of the routine maintenance, to be carried out whenever the need arises. A machine cleaned regularly will not only succumb less speedily to the inevitable corrosion of external surfaces, and hence maintain its market value, but will be far more approachable when the time comes for maintenance or service work. Furthermore, loose or failing components are more readily spotted when not partially obscured by a mantle of road grime and oil.

The moulded plastic components need to be treated in a different manner from any metal cycle parts when it comes to cleaning. They will be adversely affected by traditional cleaning and polishing techniques, and lead as a result, to the surface finish deteriorating. Avoid the use of strong detergents which contain bleaching additives, scouring powders or other abrasive cleaning agents, including all but the finest aerosol polishes. Cleaning agents with an abrasive additive will score the surface of the panels thereby making them more receptive to dirt and permanently damaging the surface finish. The most satisfactory method of cleaning the body panels is to 'float' off any dirt from their surface by washing them thoroughly with a mild solution of soapy water and then wiping them dry with a clean chamois leather. A light coat of polish may then be applied to each panel as necessary if it is thought that the panel is beginning to lose its original shine.

The plated parts of the machine should require only a wipe with a damp rag. If they are badly corroded, as may occur during the winter months, when the roads are salted, it is preferable to use one of the proprietary chrome cleaners. These often have an oily base, which will help to prevent the corrosion from recurring.

If the engine parts are particularly oily, use a cleaning compound such as 'Gunk' or 'Jizer'. Apply the compound whilst the parts are dry and work it in with a brush so that it has the opportunity to penetrate the film of grease and oil. Finish off by washing down liberally with plenty of water, taking care that it does not enter the carburettor or the electrics.

Whenever possible, the machine should be wiped down after it has been used in the wet, so that it is not garaged under damp conditions which will promote rusting. Remember there is little chance of water entering the control cables and causing stiffness of operation if they are lubricated regularly as recommended in the preceding Sections of this Chapter.

Chapter 1 Engine and transmission

Contents

Specifications

Engine	T50	T80
Type	Air-cooled, single cylinder, sohc, four-stroke	
Bore	39.0 mm (1.54 in)	47.0 mm (1.85 in)
Stroke	41.4 mm (1.63 in)	45.6 mm (1.80 in)
Displacement	49 cc (2.9 cu in)	79 cc (4.8 cu in)
Compression ratio	10.3 : 1	9.6 : 1

Cylinder head

Type	Aluminium alloy
Maximum warpage	0.5 mm (0.002 in)

Camshaft and rocker arms

	T50	T80
Cam chain type	DID25	
Number of links	82	
Cam lobe overall height:	**T50**	**T80**
Inlet	25.255 – 25.355 mm	25.257 – 25.357 mm
Service limit	25.235 mm	25.227 mm
Exhaust	25.256 – 25.356 mm	25.258 – 25.358 mm
Service limit	25.226 mm	25.228 mm
Cam base circle diameter:		
Inlet	20.994 – 21.059 mm	21.001 – 21.101 mm
Service limit	20.964 mm	20.971 mm
Exhaust	21.021 – 21.121 mm	21.025 – 21.125 mm
Service limit	20.991 mm	20.995 mm
Cam lift:		
Inlet	4.245 – 4.365 mm	4.247 – 4.367 mm
Service limit	4.227 mm	4.227 mm
Exhaust	4.246 – 4.366 mm	4.248 – 4.368 mm
Service limit	4.226 mm	4.248 mm
Valve timing (T80):		
Inlet opens at	35° BTDC	
Inlet closes at	45° ABDC	
Exhaust opens at	49° BBDC	
Exhaust closes at	31° ATDC	
Valve overlap	66°	
Rocker shaft OD	9.979 – 9.990 mm (0.3929 – 0.3933 in)	
Service limit	Not available	
Rocker arm ID	9.981 – 9.991 mm (0.3930 – 0.3934 in)	
Service limit	Not available	
Rocker arm to shaft clearance	0.009 – 0.034 mm (0.0004 – 0.0013 in)	
Service limit	0.08 mm (0.003 in)	

Valves and springs

Valve spring free length:	
Inlet and exhaust	27.3 mm (1.075 in)
Service limit	25.4 mm (1.00 in)
Valve spring warpage limit – inlet and exhaust	1.2 mm (0.047 in)
Valve clearances (cold):	
Inlet	0.050 – 0.100 mm (0.0020 – 0.0040 in)
Exhaust	0.075 – 0.125 mm (0.0030 – 0.0050 in)
Valve face contact width:	
Inlet and exhaust	0.9 – 1.1 mm (0.0354 – 0.0433 in)
Service limit	1.6 mm (0.0630 in)
Valve stem OD:	
Inlet	4.975 – 4.990 mm (0.196 – 0.197 in)
Service limit	4.950 mm (0.195 in)
Exhaust	4.960 – 4.975 mm (0.195 – 0.196 in)
Service limit	4.953 mm (0.195 in)
Valve guide ID:	
Inlet and exhaust	5.000 – 5.012 mm (0.1969 – 0.1973 in)
Service limit	5.030 mm (0.1980 in)
Valve stem to guide clearance:	
Inlet	0.010 – 0.037 mm (0.0004 – 0.0015 in)
Service limit	0.080 mm (0.003 in)
Exhaust	0.025 – 0.052 mm (0.0010 – 0.0020 in)
Service limit	0.10 mm (0.004 in)
Valve stem runout (maximum) – inlet and exhaust	0.02 mm (0.0008 in)

Cylinder barrel

Standard bore diameter:	
T50	39.000 – 39.005 mm (1.5354 – 1.5356 in)
T80	47.0 mm (1.85 in)
Taper limit	0.05 mm (0.0020 in)
Maximum ovality	0.01 mm (0.0004 in)

Piston

Diameter at base of skirt:	
T50	38.960 – 38.975 mm (1.534 in)
T80	46.975 – 46.980 mm (1.849 – 1.850 in)
Service limit	Not available
Cylinder to piston clearance	0.025 – 0.045 mm (0.00098 – 0.00177 in)
Service limit	0.1 mm (0.004 in)
Gudgeon pin bore ID	13.002 – 13.013 mm (0.5119 – 0.5123 in)
Service limit	13.055 mm (0.514 in)
Gudgeon pin OD	12.996 – 13.000 mm (0.5117 – 0.5118 in)
Service limit	12.98 mm (0.5110 in)

Piston rings

End gap (installed):	
Top and 2nd (T50)	0.08 – 0.20 mm (0.003 – 0.008 in)
Top and 2nd (T80)	0.10 – 0.25 mm (0.004 – 0.010 in)
Service limit	0.4 mm (0.016 in)
Oil	0.2 – 0.7 mm (0.008 – 0.028 in)
Service limit	N/Av
Ring to groove clearance:	
Top (T50)	0.030 – 0.065 mm (0.0012 – 0.0026 in)
Top (T80)	0.040 – 0.065 mm (0.0016 – 0.0026 in)
Service limit	0.12 mm (0.005 in)
2nd	0.02 – 0.055 mm (0.0008 – 0.0022 in)
Service limit	0.12 mm (0.005 in)

Crankshaft

Big-end bearing clearances:	
Axial	0.100 – 0.400 mm (0.0039 – 0.0157 in)
Service limit	0.50 mm (0.020 in)
Radial	0.004 – 0.019 mm (0.00016 – 0.00075 in)
Service limit	Not available
Connecting rod free play at eye end:	
Standard	0.8 – 1.0 mm (0.032 – 0.039 in)
Service limit	1.5 mm (0.059 in)
Crankshaft runout (max)	0.05 mm (0.002 in)

Clutch

Type	Automatic, centrifugal type with manual override connected to gearchange mechanism
Number of friction plates	5
Friction plate thickness	2.9 – 3.1 mm (0.114 – 0.122 in)
Service limit	2.9 mm (0.114 in)
Number of plain plates	5
Plain plate thickness	1.2 – 1.6 mm (0.047 – 0.063 in)
Plain plate maximum warpage	0.06 mm (0.0024 in)
Clutch spring free length	30.1 mm (1.185 in)
Service limit	29.1 mm (1.15 in)
Clutch commences engagement at	2600 – 2800 rpm (T50), 2100 – 2300 rpm (T80)
Clutch fully engaged at	3400 – 3600 rpm (T50), 2800 – 3000 rpm (T80)
Manual control	Pushrod operated by gearchange pedal

Transmission

Primary drive	Helical gear	
Primary reduction ratio – T50	3.722 : 1 (67/18T)	
Primary reduction ratio – T80	3.250 : 1 (65/20T)	
Gearbox type	3-speed (T50) or 4-speed (T80) constant mesh	
Gear ratios:	**T50**	**T80**
1st	3.250 : 1 (39/12T)	3.545 : 1 (39/11T)
2nd	1.812 : 1 (29/16T)	1.888 : 1 (34/18T)
3rd	1.200 : 1 (24/20T)	1.260 : 1 (29/23T)
4th	Not applicable	1.040 : 1 (26/25T)
Secondary drive	Shaft	
Secondary drive ratio – T50	3.589 : 1 (19/18 x 34/10)	
Secondary drive ratio – T80	3.166 : 1 (19/18 x 33/11)	
Gearbox input shaft runout – service limit	0.08 mm (0.0031 in)	
Gearchange mechanism type	Drum	
Selector fork claw thickness (T80):		
Standard	4.76 – 4.89 mm (0.187 – 0.193 in)	
Service limit	Not available	
Kickstart mechanism type	Ratchet	
Kickstart friction clip force	0.8 – 1.5 kg (1.76 – 3.31 lb)	

Torque wrench settings

Component	kgf m	lbf ft
Cylinder head and barrel – 6 mm nut and 6 mm bolt	1.0	7.2
Cylinder head cover	0.7	5.1
Spark plug	1.25	9.0
Alternator rotor nut	4.0	29.0
Valve adjuster locknut	0.7	5.1
Cam sprocket bolt	2.0	14.0
Cam chain tensioner bolt	1.5	11.0
Tensioner locknut	0.7	5.1
Tensioner top nut – T50	0.5	3.6
Tensioner top nut – T80	0.7	5.1
Oil pump cover screws	0.4	2.9
Oil pump mounting bolts	0.7	5.1
Crankcase drain plug	2.0	14.0

Component	kgf m	lbf ft
Inlet adaptor	0.7	5.1
Carburettor mounting	0.7	5.1
Exhaust port nuts	1.0	7.2
Crankcase fasteners:		
6 mm screws	0.7	5.1
6 mm bolts	0.7	5.1
Kickstart lever pinch bolt	1.2	8.7
Primary drive gear nut	5.0	36.0
Clutch release lever pivot	0.8	5.8
Clutch centre nut - T50	5.5	40.0
Clutch centre nut - T80	5.0	36.0
Front bevel gearbox nuts	6.0	43.0
Driven gear nut	9.0	65.0
Driven gear housing	1.0	7.2
Gearchange stopper plate	0.8	5.8
Gearchange stopper lever	1.4	10.0
Gearchange pedal pinch bolt	2.5	18.0
Driveshaft housing nut	6.0	43.0
Driveshaft bearing housing bolts	1.0	7.2
Ignition pickup coil	0.4	2.9
Alternator stator	0.7	5.1
Neutral switch	1.0	7.2

1 General description

The Yamaha T50 and 80 models employ the same basic design of single-cylinder, sohc (single overhead camshaft) four-stroke engine as that used on successive earlier step-through models produced by the various Japanese motorcycle manufacturers. The engine's single cast iron cylinder is inclined forward to allow the frame to loop down between the steering head and seat, and is partially shrouded by the moulded plastic legshield. The light alloy cylinder head houses the camshaft and valve gear, the camshaft drive being by a single-row roller chain on the left-hand side of the unit. Chain tension is controlled semi-automatically.

Power from the engine unit is fed to a conventional foot-controlled four-speed gearbox. An automatic centrifugal clutch is designed to engage as the engine speed rises above about 2100 rpm. In addition, a linkage from the gearchange pedal disengages the clutch when the pedal is lifted or depressed. This arrangement dispenses with the need for a clutch lever, and limits the rider's hand-operated controls, to a throttle twistgrip and the front brake lever.

2 Operations with the engine/transmission unit in the frame

Many of the engine/transmission components and assemblies can be worked on with the unit installed in the frame, the main items being listed below. Note, however, that it is normal practice to remove the unit where all but the simplest overhauls is being contemplated; the unit sits low in the frame and is tiring to work on unless the whole machine can be raised to a convenient working height. Engine/transmission unit removal takes only a few minutes and provides much better access to all areas. If necessary, however, the following items can be worked on with the engine unit in the frame:

(a) The cylinder head, valves and camshaft
(b) The cylinder barrel and piston
(c) The alternator
(d) The transmission clutch and external gearchange components
(e) The oil pump
(f) The kickstart mechanism

3 Operations requiring engine/transmission unit removal

The following components and assemblies can only be reached after removing the engine/transmission unit from the frame, dismantling the components listed in the previous Section, and separating the crankcase halves:

(a) The crankshaft assembly
(b) The gearbox shaft assemblies
(c) The crankshaft and transmission shaft bearings and bushes
(d) The kickstart shaft internal components

4 Removing the engine/transmission unit from the frame

1 As has been mentioned earlier in this Chapter, the procedure for removing the engine/transmission unit is quite straightforward, and with a little experience should take only a matter of minutes to complete. For this reason it is recommended that the engine is removed before undertaking all but the simplest jobs; the improved access and working position is well worth the extra time involved, and the unit can be properly cleaned before dismantling begins.

2 Before engine/transmission removal is commenced, it is necessary to remove the moulded plastic legshield assembly. Note that this is normally required even if the work is to be carried out with the unit in the frame. The legshield assembly is retained by a total of six cross-head screws. Remove the screws, then manoeuvre the moulding clear of the frame.

3 Care should be taken when handling the legshield moulding; it is resilient, but is easily scratched. Place it on soft cloth in a safe place when removed, and take care not to score the surface during removal or fitting.

4 Place a drain tray of about two Imp pint (1 litre) capacity below the crankcase, unscrew the drain plug, and allow the engine/transmission oil to drain. Leave the oil to drain thoroughly before cleaning the drain plug and its thread and refitting the plug, using a new sealing washer if necessary. The plug should be tightened to 2.0 kgf m (14 lbf ft). While the oil is draining, remove the right-hand side panel and disconnect the battery leads (negative lead first). This will isolate the electrical system to prevent accidental short circuits.

5 Slacken and remove the two nuts which retain the exhaust pipe at the cylinder head, then free the two silencer mounting nuts. The latter are located on the inside edge of the silencer, and can be reached using an open ended or combination spanner. The complete exhaust system can now be lifted away. Remove the bolts which retain the footrest assembly and lift it clear of the underside of the engine unit.

6 Remove the gearchange pedal pinch bolt, then slide the pedal off its splines. Remove the screws which secure the crankcase left-hand outer cover and lift it away. Remove the left-hand side panel, then trace back the ignition pickup and alternator wiring, separating them at the connectors located behind the panel. Work off the spring retainer which secures the gaiter from the driveshaft outer casing to the corresponding section of the crankcase unit, then pull the gaiter free of the crankcase. The gaiter will remain attached to the driveshaft casing when the engine unit is removed. Disconnect the spark plug lead at the plug, lodging the lead and cap clear of the engine unit.

7 Moving to the carburettor, disconnect the throttle and choke cables, lodging them clear of the carburettor and engine unit. Slacken the retaining clip which secures the air filter inlet hose to the

carburettor mouth, and pull the hose free of the carburettor. Check that the fuel tap is set to the 'ON' position, then disconnect the fuel and vacuum pipes. (Note that if the tap is set to the 'PRI' position, fuel will run out from the tank when the pipe is disconnected.)
8 On the right-hand side of the unit, disconnect the rear brake pedal return spring, allowing the pedal to drop slightly. This will provide additional clearance during engine removal. The unit is now ready to be removed from the frame. Before releasing the mounting bolts, check around for any missed pipes or electrical leads which might impede removal. The engine unit is not particularly heavy, and being compact is easily removed by one fairly strong person. It is preferable, however, to have a second pair of hands available to help steady the machine and to withdraw the mounting bolts.
9 Remove the two mounting bolt nuts, then supporting the weight of the engine unit, displace and remove the mounting bolts. Lower the unit clear of the frame, manoeuvring it forward to allow the driveshaft coupling to slide apart. Place the assembly on the workbench to await further dismantling.

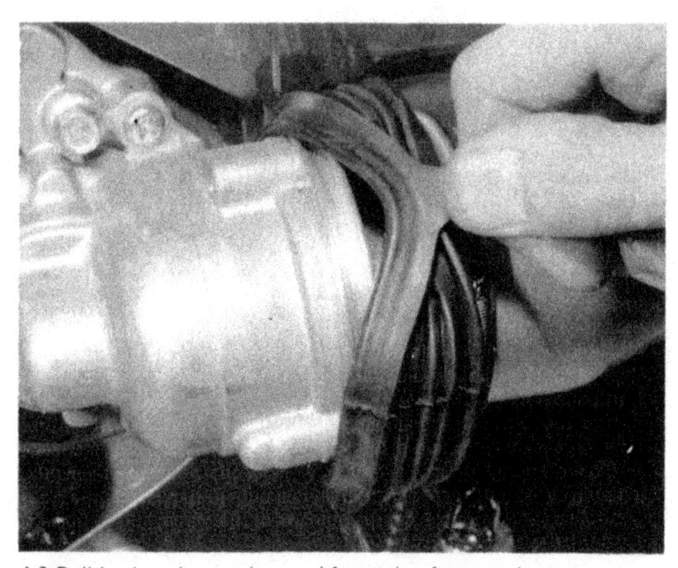

4.6 Roll back spring retainer and free gaiter from crankcase

4.7 Disconnect cables, fuel and vacuum pipes at carburettor

5 Dismantling the engine/transmission unit: general

1 Before any dismantling work is undertaken, the external surfaces of the unit should be thoroughly cleaned and degreased. This will prevent the contamination of the engine internals, and will also make working a lot easier and cleaner. A high flash-point solvent, such as paraffin (kerosene) can be used, or better still, a proprietary engine degreaser such as Gunk. Use old paintbrushes and toothbrushes to work the solvent into the various recesses of the engine castings. Take care to exclude solvent or water from the electrical components and inlet and exhaust ports. The use of petrol (gasoline) as a cleaning medium should be avoided, because the vapour is explosive and can be toxic if used in a confined space.
2 When clean and dry, arrange the unit on the workbench, leaving a suitable clear area for working. Gather a selection of small containers and plastic bags so that parts can be grouped together in an easily definable manner. Some paper and a pen should be on hand to permit notes to be made and labels attached where necessary. A supply of clean rag is also required.
3 Before commencing work, read through the appropriate section so that some idea of the necessary procedure can be gained. When removing the various engine components it should be noted that great force is seldom required, unless specified. In many cases, a component's reluctance to be removed is indicative of an incorrect approach or removal method. If in any doubt, re-check with the text.

6 Dismantling the engine/transmission unit: removing the cylinder head

1 If the cylinder head is to be removed with the engine/transmission unit installed in the frame, the legshields, exhaust system and carburettor assembly should be removed before proceeding further, this procedure being described in Section 4. It will also be necessary to drain the engine/transmission oil and to remove the gearchange pedal and the crankcase left-hand outer cover.
2 The carburettor and the inlet adaptor should be removed from the inlet port. If the engine unit has been removed from the frame, the control cables, air filter hose and the fuel and vacuum pipes will already have been dealt with. With the unit in the frame, release the air filter hose retaining clip. The remaining connections can be left in place and the assembly tied clear of the cylinder head, although if required they can be disconnected and the carburettor lifted away to gain better access. Once again, this is described in Section 4 of this Chapter.
3 Slacken and remove the spark plug so that the crankshaft can be turned easily. Using the alternator rotor, turn the crankshaft anticlockwise until the 'T' mark on the alternator rotor aligns with its fixed index mark on the crankcase. Remove the circular end cover on the left-hand end of the cylinder head (two screws) then release the tensioner mechanism which is located immediately above it as follows: Remove the domed nut, slacken the locknut, then unscrew the small locking screw. At the top of the cylinder head casting, remove the tensioner end plug, spring and plunger.
4 Using a fabricated rotor holding tool (see accompanying line drawing) or a strap wrench, hold the alternator rotor to prevent the crankshaft from turning. Slacken and remove the cam sprocket bolt, then lift away the sprocket. Take care not to allow the cam chain to drop down into its tunnel; a length of wire can be used to secure the chain and prevent this from happening.
5 Remove in a diagonal sequence the four domed nuts which retain the cylinder head. Remove the two 6 mm bolts which retain the cylinder head to the barrel. The cylinder head should now lift away from the barrel. If it seems stuck, try tapping around the joint with a soft-faced hammer to break the seal.
6 If this fails, try placing a hardwood block against the underside of the cylinder head and striking it sharply to separate the joint. **On no account** attempt to lever between the head and barrel. Place the head to one side and then remove the dowel pins (if they are loose) and the head gasket. These will probably have remained attached to the upper face of the cylinder barrel. The cam chain tensioner guides can be left in place unless they require attention.

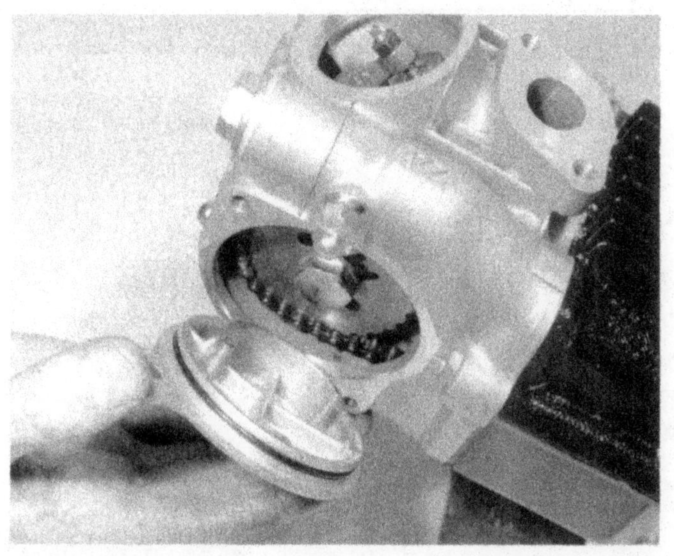

6.4 Remove inspection cover and release single bolt to free the cam sprocket

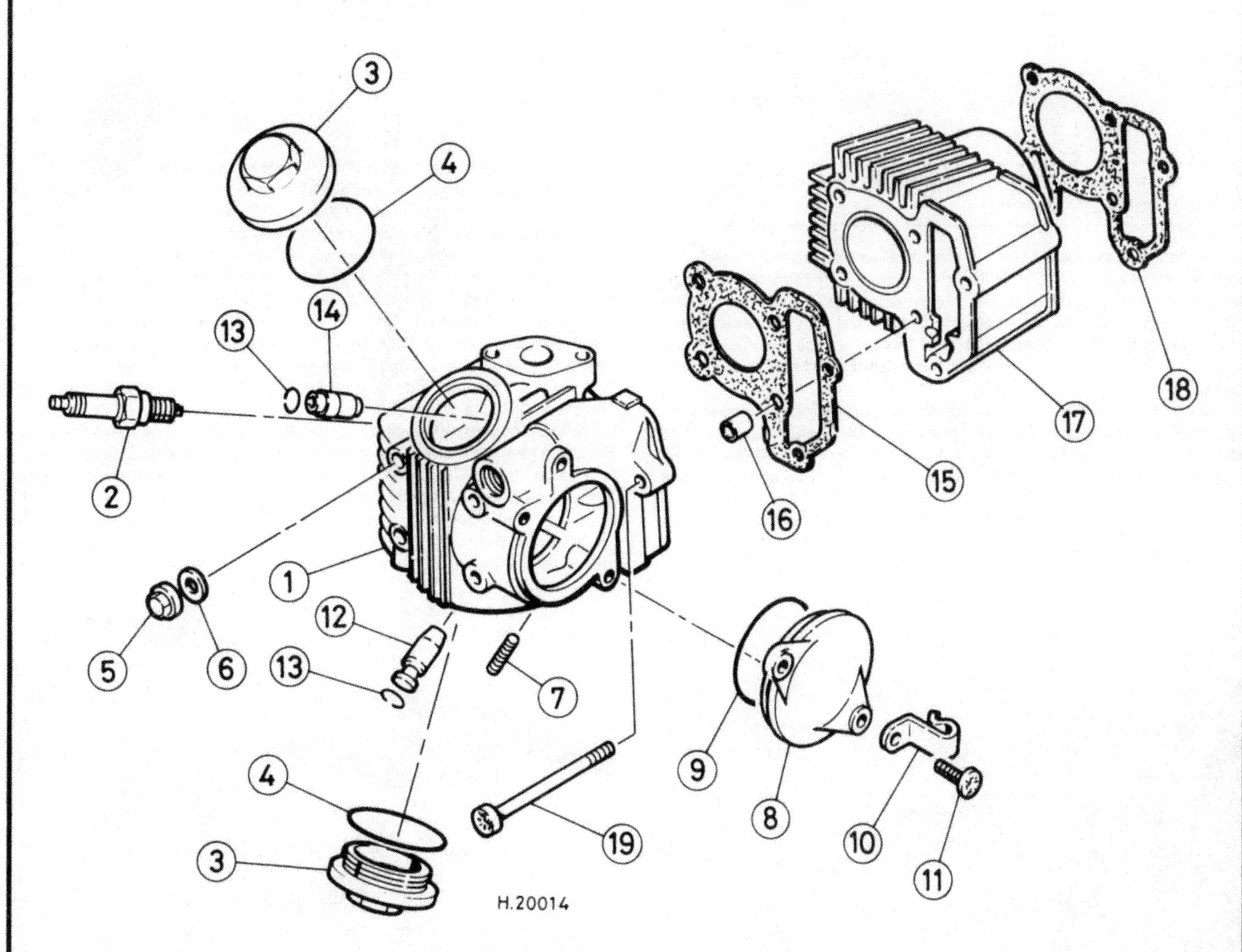

Fig. 1.1 Cylinder head and barrel

1 Cylinder head
2 Spark plug
3 Valve clearance inspection cap – 2 off
4 O-ring – 2 off
5 Nut – 4 off
6 Washer – 4 off
7 Stud – 2 off
8 Cylinder head side cover
9 O-ring
10 Cable clamp
11 Screw – 2 off
12 Exhaust valve guide
13 Clip – 2 off
14 Inlet valve guide
15 Cylinder head gasket
16 Dowel pin – 2 off
17 Cylinder barrel
18 Cylinder base gasket
19 Bolt – 2 off

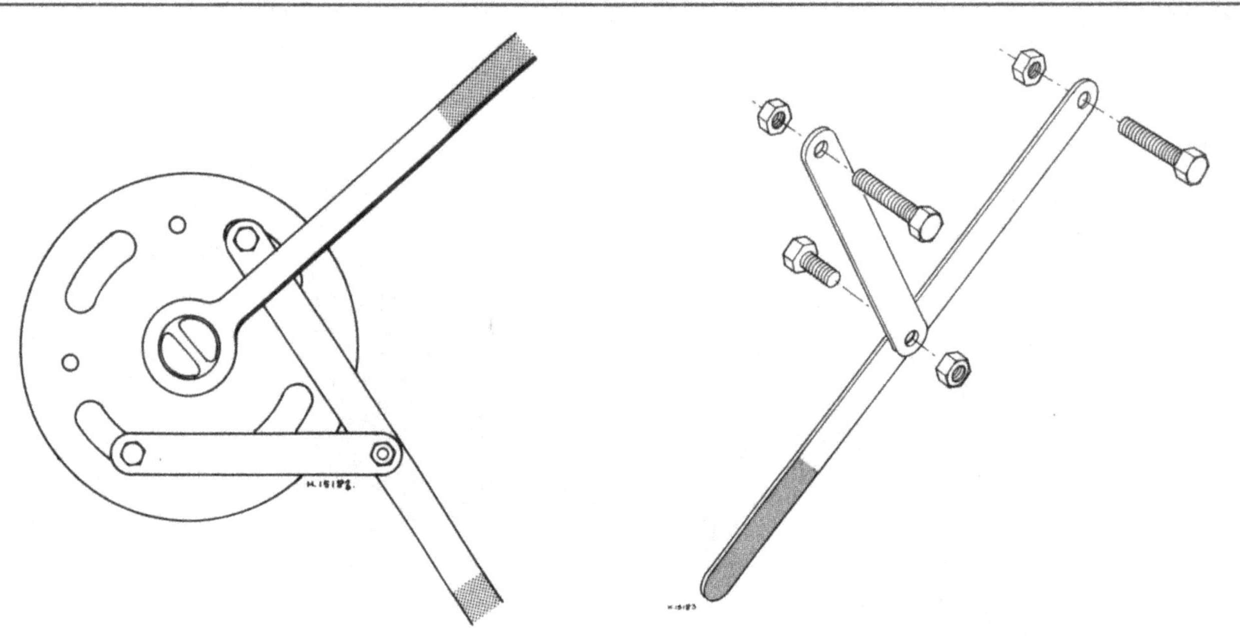

Fig. 1.2 Fabricated alternator rotor holding tool

Method of use *Construction of tool*

7 Dismantling the engine/transmission unit: removing the cylinder barrel and piston

1 If the cylinder barrel and piston are to be removed with the engine/transmission unit installed in the frame, the legshields, exhaust system and carburettor assembly should be removed before proceeding further. It will also be necessary to drain the engine/transmission oil, remove the gearchange pedal and the crankcase left-hand outer cover, and to remove the cylinder head as described in the preceding Sections.

2 With the cylinder head removed, disengage and remove the tensioner blade and guide, making a sketch of their relative positions to aid reassembly. The cylinder barrel is now free to be lifted away from the crankcase. If this proves difficult, it can be dislodged using either of the methods described above for the cylinder head.

3 Some care should be taken to avoid any debris from dropping down into the crankcase; if this should happen the crankcase halves must be separated to allow them to be cleaned out. The cylinder angle of the T80 models minimises this risk, but to be sure, some rag should be packed into the crankcase mouth **before** the piston leaves the bore completely. Continue to slide the barrel off the piston, supporting the latter as it emerges from the bottom of the bore.

4 To free the piston from the connecting rod, prise out one of the circlips, using fine pointed-nosed pliers or a small screwdriver via the slot provided. Take great care to avoid scoring the piston during this operation. Push the gudgeon pin through from the other side until it clears the small-end bush, then lift the piston away. Finger pressure is normally sufficient to move the gudgeon pin, but if it proves stiff, check that it is not being held by a build-up of carbon at the end of the gudgeon pin bore or around the circlip groove.

5 If it still fails to move, support the piston firmly in one hand to avoid placing a strain on it or the connecting rod, then use a double-diameter drift to tap the pin out. **Do not** use excessive force. If the pin proves abnormally tight, expand the light alloy piston to release its grip on the pin by wrapping some rag soaked in nearly boiling water around it for a few minutes. Note that care must be taken to avoid scalding the hands; use some industrial rubber gloves or similar to prevent this possibility.

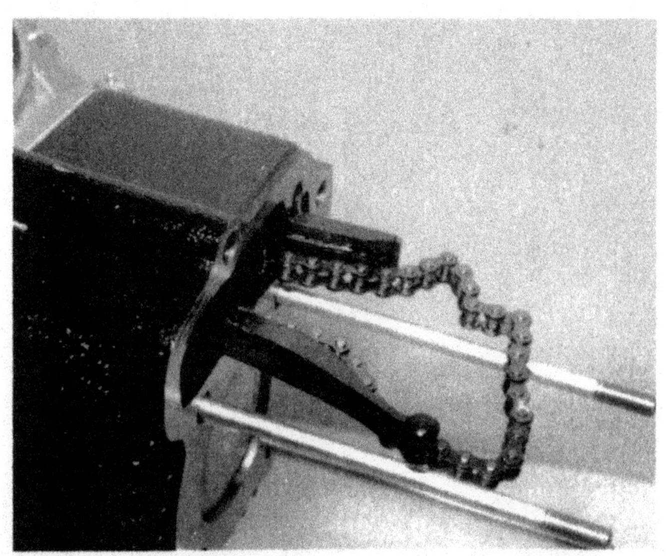

7.2 Remove the tensioner blades and slide barrel off its studs. Support piston as it emerges from bore

7.4 Prise out circlip, then displace gudgeon pin to free the piston

8 Dismantling the engine/transmission unit: removing the alternator and cam chain

1 If the alternator is to be removed with the engine/transmission unit installed in the frame, the gearchange pedal and the crankcase left-hand outer cover should be removed before proceeding further. Note also that if the alternator stator is to be removed, the engine/transmission oil must first be drained; the stator casting closes the left-hand side of the crankcase.
2 Removal of the rotor will require an extractor, available either as a Yamaha Special Tool through Yamaha dealers, or as a pattern tool from most motorcycle dealers. Attempting to remove the rotor without this tool is not recommended; there is a serious risk of damage by using other methods.
3 The rotor must be held while the retaining nut is slackened, and Yamaha produce a holding tool which can be ordered through official dealers. A similar tool is easily fabricated using steel strip, see Fig. 1.2. In the absence of the correct tool, a strap or chain wrench can be used, provided that great care is taken to avoid damage to the external ignition pickup. Alternatively, if the cylinder head, barrel and piston have been removed, the crankshaft (and thus the rotor) can be locked by passing a smooth metal bar through the connecting rod small-end eye and supporting its ends on strips of wood placed on each side of the crankcase mouth.
4 With the rotor immobilised, slacken and remove the retaining nut. Screw the extractor into the rotor boss **(left-hand thread),** until it seats fully. Tighten the extractor centre bolt to draw the rotor off its taper. If it seems tight, **do not** continue tightening; tap the end of the extractor centre bolt sharply to jar the rotor free. Lift the rotor away and place it to one side, together with the Woodruff key if this drops out of its slot in the crankshaft.
5 Remove the left-hand side panel and trace back the ignition pickup and alternator wiring, separating it at the connectors on the CDI unit and regulator/rectifier unit. The stator assembly is secured by two screws to the crankcase and may be removed once these have been unscrewed. Note that the stator and ignition pickup coils should not be disturbed. The stator is sealed into its recess by a large O-ring, and may require some judicious prising with a pair of small screwdrivers to extract it. As the stator assembly comes free from the crankcase, disconnect the neutral switch wiring at its terminal, disengage the output wiring grommet from its slot and place the assembly to one side. The cam chain can now be disengaged from the crankshaft sprocket, assuming that the cylinder head and barrel have been removed as described in the previous sections.

8.4 Use rotor extractor to draw rotor off the crankshaft taper

8.5a Remove the two screws to free the stator assembly from the crankcase

8.5b Cam chain can now be disengaged and removed

9 Dismantling the engine/transmission unit: removing the middle gear assembly

The middle gear assembly comprises an alloy bearing plate, in which is carried the middle gear shaft and universal joint. These need not be disturbed to permit crankcase separation, but if removal is deemed necessary, slacken and remove the three retaining bolts and pull the assembly clear of the crankcase. Place the assembly to one side to await further attention.

10 Dismantling the engine/transmission unit: removing the centrifugal clutch, kickstart mechanism, oil pump, gearchange shaft and primary drive gear

1 The transmission clutch may be removed with the engine/transmission unit in or out of the frame. If the unit is in the frame, it will first be necessary to drain the engine/transmission oil and to remove the exhaust system.
2 Irrespective of whether the engine is in the frame or on the workbench, it will be necessary to remove the footrest assembly, together with the side stand. This is retained by bolts to the underside

of the crankcase. The kickstart lever must also be removed, this being retained by a single clamp bolt to its splined shaft.

3 Before removing the crankcase right-hand outer cover, make a sketch or note of the position of any wiring guide clips; this will save time during reassembly. Have a drain tray handy to catch the residual oil from the cover, then slacken evenly the cover screws. Lift away the cover, together with the gasket and dowel pins.

4 With the cover removed and any excess oil drained away, attention can be given to removing the clutch. Start by removing the large internal circlip which retains the clutch pressure plate assembly. Using a drift or a similar tool, press against the pressure plate whilst working one end of the circlip out of its groove. Continue working around the circlip until it comes free. Maintain pressure on the plate assembly to avoid damaging the groove as the clip comes free and the pressure plate is displaced. Lift away the short (outer) and long (inner) clutch pushrods and place them with the pressure plate.

5 Lift away the clutch plain and friction plates as an assembly, together with the clutch springs. It is now necessary to remove the clutch centre nut, an operation that requires the clutch centre to be locked in some way. This can be done by using the official holding tool which can be ordered through Yamaha dealers, or by making up a tool such as that shown in the accompanying line drawing. As an alternative, the crankshaft may be locked by passing a smooth round bar through the connecting rod small-end eye, its ends resting on wooden strips placed against the crankcase mouth to prevent marking of the gasket face. Note that the latter method assumes removal of the cylinder head, barrel and piston has already taken place.

6 With the clutch centre immobilised, slacken and remove the retaining nut, then lift away the washer, thrust plate, wave washer, large plain washer and the one-way boss. The clutch centre can now be lifted away, followed by the clutch outer drum complete with the steel balls. The latter need not be disturbed at this stage. Remove the bearing sleeve and plain thrust washer, if these have remained on the gearbox input shaft end, and place them to one side with the rest of the clutch assembly.

7 The gearchange shaft assembly can be removed after the clutch has been dismantled, as described above. If it is still in position, the crankcase left-hand outer cover must also be removed to gain access to the clutch actuating mechanism. Working from the left-hand side of the unit, remove the circlip which secures the cam to the shaft end and slide it off. Displace the pin which passes through the shaft, then lift away the outer thrust washer, thrust bearing and inner thrust washer to free the shaft. Moving to the right-hand side of the unit, grasp the end of the gear selector arm, pulling the claw end away from the end of the selector drum. The shaft can now be withdrawn, the ends of the centring spring dropping clear of the anchor pin as it comes clear of the casing.

8 To free the kickstart mechanism, grasp the end of the return spring using a pair of pliers, and unhook it from the anchor pin on the crankcase. Allow the spring to unwind to release spring pressure. The assembly can now be removed from its recess and placed to one side. Note that it is not necessary to dismantle the assembly further unless it requires attention, this procedure being described later in this Chapter.

9 The crankshaft primary drive gear assembly can be removed after the crankshaft has been locked to prevent its rotation. Use either the alternator rotor holding tool, described earlier in this Chapter, or pass a smooth round bar through the connecting rod small-end eye and support its ends on wooden blocks arranged on each side of the crankcase mouth. Once again, the latter method presupposes that the cylinder head, barrel and piston are removed. With the crankshaft locked in position, slacken and remove the retaining nut, followed by the primary drive gear and the centrifugal filter unit.

10 Once the clutch, primary drive pinion and centrifugal filter have been detached, the oil pump can be removed, if required. Note, however, that the pump does not prevent crankcase separation, and thus can be left in position. The pump pinion should be removed first, this being retained to the pump spindle by a circlip. Prise off the clip, then lift away the pinion and thrust washer. Slacken the two cross-head screws which hold the pump body to the crankcase and lift it away. If the oil pickup strainer is still in position, this should be pulled out of its slot in the crankcase and placed to one side for cleaning.

10.4 Prise out the large circlip which retains the clutch pressure plate assembly

10.5 Lift away the clutch plates, together with the springs

10.6a Use home-made holding tool while releasing clutch centre nut

10.6b Lift away clutch centre, then outer drum and release balls

10.9a Remove nut, then lift away the primary drive pinion

10.9b The centrifugal filter can now be slid off the crankshaft

10.10a Release the oil pump pinion by removing circlip

10.10b The pump is retained by two cross-head screws

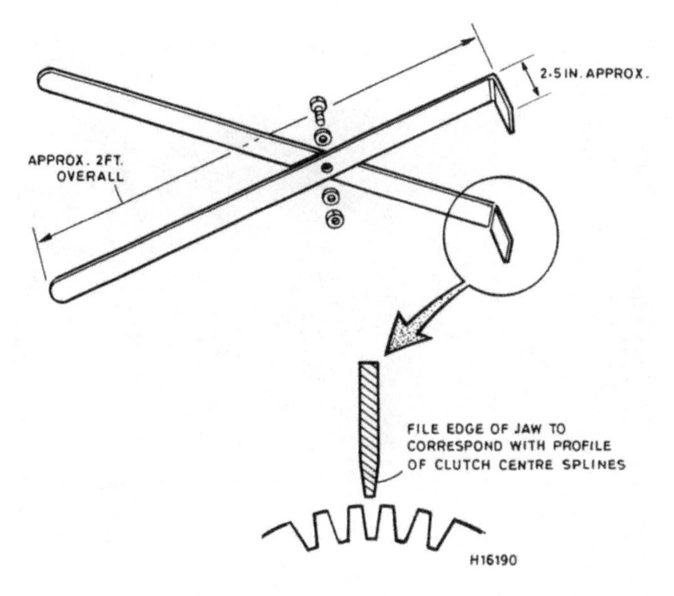

Fig. 1.3 Fabricated clutch holding tool

11 Dismantling the engine/transmission unit: separating the crankcase halves

1 Crankcase separation is a necessary to gain access to the crankshaft, gearbox shafts and the gear selector drum and forks. Before the crankcase halves can be separated it is first necessary to remove the engine/transmission unit from the frame. The cylinder head, cylinder barrel, alternator, centrifugal clutch, gearchange shaft assembly and the oil pump must then be removed as detailed in the preceding Sections.
2 Make a note of the position of all guide clips, then slacken and remove the nine screws which hold the crankcase halves together and the three bolts which retain the output shaft to the left-hand casing. Slacken the screws evenly and progressively to prevent any risk of warpage, preferably using an impact driver to minimise the risk of damage to the heads. It will also be necessary to remove the selector drum retainer, this being held by two cross-head screws on its right-hand end.
3 Support the unit on its side on wooden blocks, with the right-hand casing half uppermost. Tap around the crankcase joint with a soft-faced hammer, or with a metal hammer and a hardwood block, to assist in breaking the seal between them. To assist in separating the crankcase halves, pry points are cast into the casing at the front and rear, and a pair of screwdrivers can be used to gently ease the casing halves apart.
4 The right-hand casing half should now lift away from the left-hand half, but if it is securely stuck, try tapping the right-hand casing upwards, using a wooden block as a drift. With persistence this will succeed in breaking the bond of the gasket cement, and once started it is quite easy to work the cases apart. On no account resort to levering between the gasket faces; this will only result in damaged surfaces and a leaking joint after reassembly.
5 With the crankcase right-hand half removed, peel off the gasket and remove the dowel pins. The crankshaft should have remained in the left-hand half, and this can now be removed and placed to one side. The transmission input and output shaft assemblies can now be lifted out of the casing, along with the gear selector drum and forks. Place them to one side to await further attention.

12 Examination and renovation: general

1 Before examining the parts of the dismantled engine unit for wear it is essential that they should be cleaned thoroughly. Use a petrol/paraffin mix or a high flash-point solvent to remove all traces of old oil and sludge which may have accumulated within the engine. Where petrol is included in the cleaning agent normal fire precautions should be taken and cleaning should be carried out in a well ventilated place.
2 Examine the crankcase castings for cracks or other signs of damage. If a crack is discovered it will require a specialist repair.
3 Examine carefully each part to determine the extent of wear, checking with the tolerance figures listed in the Specifications section of this Chapter or in the main text. If there is any doubt about the condition of a particular component, play safe and renew.
4 Use a clean lint-free rag for cleaning and drying the various components. This will obviate the risk of small particles obstructing the internal oilways, and causing the lubrication system to fail.
5 Various instruments for measuring wear are required, including a vernier gauge or external micrometer, an internal micrometer and a set of standard feeler gauges. Additionally, although not absolutely necessary, a dial gauge and mounting bracket are invaluable for accurate measurement of endfloat, and play between components of very low diameter bores – where a micrometer cannot reach. After some experience has been gained the state of wear of many components can be determined visually or by feel and thus a decision on their suitability for continued service can be made without resorting to direct measurement.

13 Examination and renovation: crankshaft assembly

1 Failure of either the big-end or main bearings is normally obvious before the engine unit is stripped. In the case of the big-end bearing, a noticeable knocking noise, usually growing worse when under load, will be heard. There is also likely to be an increase in engine vibration. Main bearing failure is characterised by a rumbling noise, again accompanied by increased engine vibration. Failure of any of the engine bearings is often due to damage from dirty oil, a reminder of the need for regular oil changes.
2 There must be no detectable radial (up-and-down) play in the big-end bearing. This can be checked by grasping the connecting rod and attempting to move it in relation to the crankpin on which it runs. It is difficult to assess the degree of wear without a set of V-blocks and a dial gauge; equipment unlikely to be available to most owners. The method recommended by Yamaha for checking big-end bearing wear is to grasp the connecting rod at the small-end eye and to attempt to rock it from side to side. Take great care when using this method not to confuse the normal side-to-side clearance (axial play) of the big-end bearing with bearing wear. If movement exceeds 0.8 – 1.0 mm (0.0315 – 0.0394 in), the big-end bearing will require attention. An authorized Yamaha dealer or engineering company will be able to confirm the diagnosis if required, although if the bearing has 'run' it is normally quite obvious.
3 The big-end bearing axial play (endfloat) should also be checked. Unlike radial play, axial play is both intentional and necessary. The prescribed minimum clearance (service limit) is 0.10 mm (0.00394 in), whilst the maximum figure is 0.50 mm (0.0197 in). This can be checked using feeler gauges between the connecting rod big-end eye

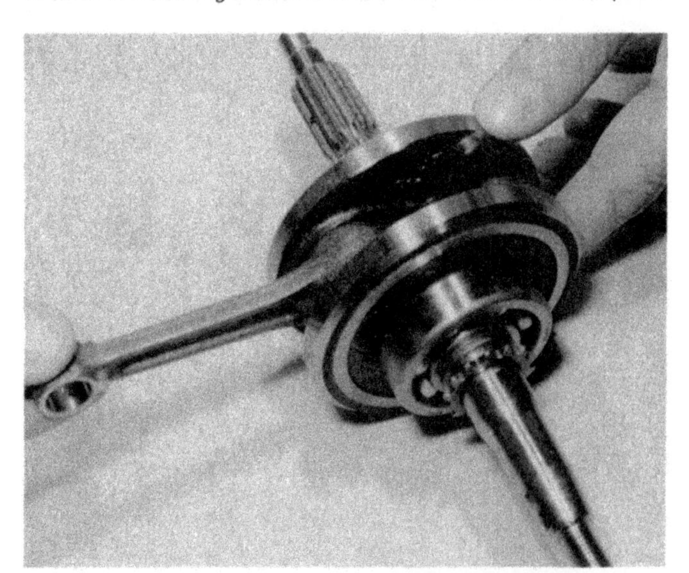
13.2 Check the big-end bearing for radial play as shown

13.4 Check the condition of the left-hand main bearing, and its fit on the crankshaft

and the adjacent flywheel face. It should be noted that any defect in the big-end calls for specialist attention. Most Yamaha dealers will be able to carry out the required work, or will be able to arrange this service with a local engineering works. The job of dismantling, overhauling, assembling and trueing the flywheel assembly is a skilled one, and should not be attempted without the necessary experience and equipment.

4 The crankshaft main bearings are of the journal ball type, and are best checked by feel; it is not easy to measure wear in bearings of this type. Spin each bearing by hand, feeling for any roughness or tight spots as the outer race turns. If the bearing feels rough, or fails to turn smoothly and evenly, it is a good indication of wear and the need for renewal. The fit of the left-hand main bearing on the projecting end of the crankshaft should also be checked. If the bearing inner race turns or feels loose on the shaft, the bearing must be renewed. In extreme cases, the shaft may also have become worn. There are a number of ways of dealing with this problem without resorting to a new crankshaft, but expert advice should be sought.

5 To remove the bearing from the crankshaft end, a bearing extractor should be used. These can be hired from tool hire shops, or alternatively, an authorized Yamaha dealer should be able to remove the old bearing and fit a new one for you. The procedure for removing the bearing, together with the camshaft drive sprocket is discussed in Section 14. The right-hand main bearing will normally have remained in the crankcase during separation, but can be checked in much the same way. If renewal is required, it is best to warm the crankcase half in near boiling water before attempting removal. This will expand the alloy casing, making removal much easier. Tap out the old bearing squarely, using a drift. The new bearing can be fitted in the same way, but drive it home hitting the **outer** race only to avoid damage. Ensure that the bearing seats squarely and fully in its recess.

6 Finally, if V-blocks and a dial gauge are available, the crankshaft runout can be checked. Set the crankshaft with its main bearings supported on the V-blocks, and arrange the dial gauge so that the probe tip rests against each of the mainshaft ends in turn. Rotate the crankshaft and note the runout indicated on the gauge. The service limit is 0.05 mm (0.002 in). Correcting crankshaft runout, like crankshaft reconditioning, is a skilled operation, and it is all too easy to make matters worse if trueing is attempted at home. Take the assembly to an authorized Yamaha dealer or a local specialist for this work to be carried out.

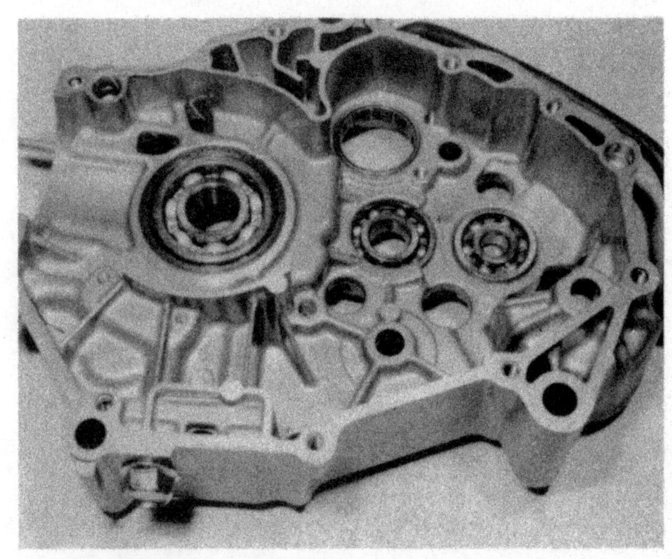

13.5 The right-hand main bearing will have remained in place in its crankcase half

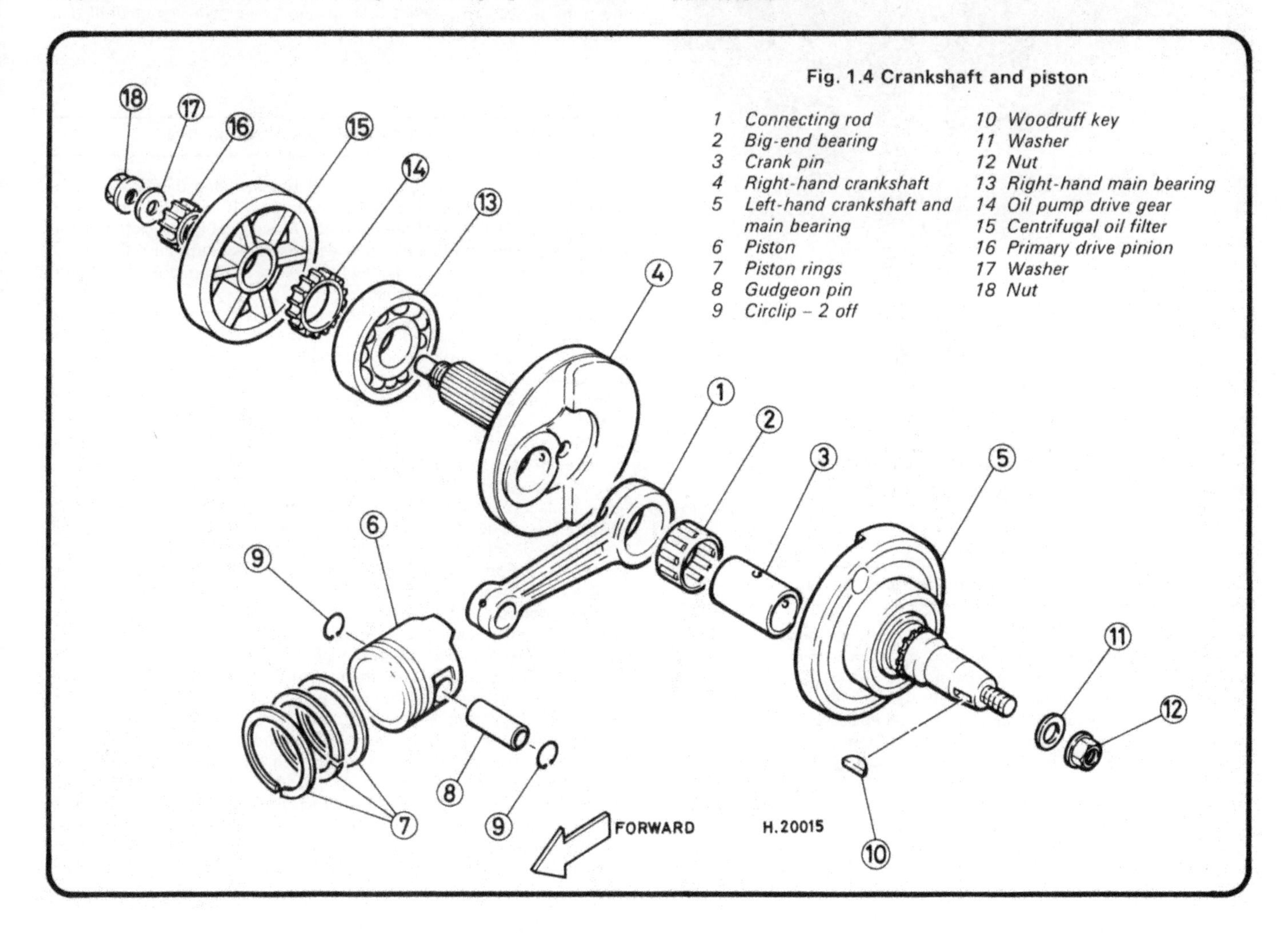

Fig. 1.4 Crankshaft and piston

1 *Connecting rod*
2 *Big-end bearing*
3 *Crank pin*
4 *Right-hand crankshaft*
5 *Left-hand crankshaft and main bearing*
6 *Piston*
7 *Piston rings*
8 *Gudgeon pin*
9 *Circlip – 2 off*
10 *Woodruff key*
11 *Washer*
12 *Nut*
13 *Right-hand main bearing*
14 *Oil pump drive gear*
15 *Centrifugal oil filter*
16 *Primary drive pinion*
17 *Washer*
18 *Nut*

14 Examination and renovation: removing and refitting the left-hand main bearing and camshaft drive sprocket

1 If examination of the left-hand main bearing has shown it to be in need of renewal, or if the camshaft drive sprocket needs to be renewed, some means of drawing them off the crankshaft will be needed. This is best done using a bearing extractor. These can be hired from tool hire specialists, or alternatively, a local garage may be able to carry out the job for you. On no account try driving off the bearings or sprockets by using wedges, or a lever between them and the flywheel; this will only distort the crankshaft assembly.
2 The new bearing can be tapped into place using a length of tubing as a drift. To prevent distortion, place a wedge between the flywheels opposite the crankpin, and support the assembly on its side on stout blocks, making sure that the right-hand crankshaft end is well clear of the bench. Tap the bearing home squarely, using no more than the minimum of force. The camshaft drive sprocket is fitted in a similar fashion, but note that it is essential that it is aligned correctly.
3 Using a fine spirit-based marker, draw a line down the crankshaft end, passing through the centre of the rotor keyway. It is essential that the line is square in relation to the shaft. Apply a thin film of medium strength locking fluid, such as Loctite 641 Bearing Fit, to the crankshaft where the sprocket will fit. (**Do not** use a permanent, high-strength compound.) Offer up the sprocket, aligning the index mark on the sprocket with the reference line. Tap the sprocket squarely down the shaft as described above, making absolutely certain that it remains in alignment. Note that if the sprocket is fitted incorrectly, it will be impossible to set the valve timing accurately.

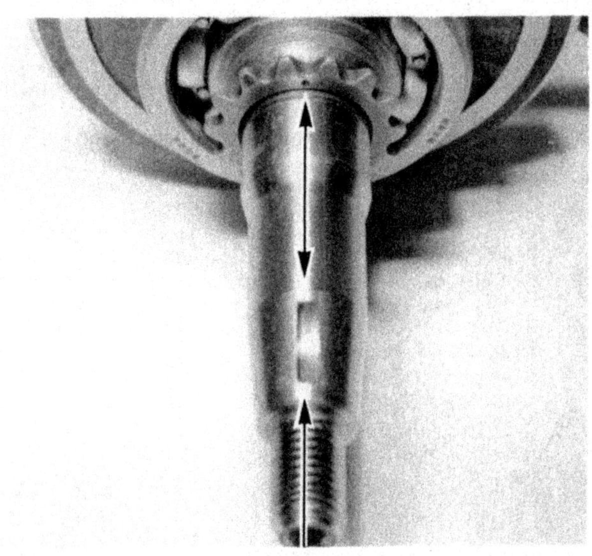

14.3 Sprocket alignment mark must coincide with the crankshaft keyway as shown

15 Examination and renovation: cylinder barrel

1 Inspect the surface of the cylinder bore for signs of wear or damage. Scoring along the length of the bore surface indicates the need for reboring and a new piston, and if such damage is noted, further inspection will be unnecessary. Only if the scoring takes the form of very light scratching can the bore be re-used. In such cases it is preferable to have the scratches removed by honing the bore, a service that most engine reconditioning companies will be able to offer. Note that if honing takes the bore beyond its maximum diameter it will be necessary to bore to the next oversize anyway, so check the existing bore size before having honing work done.
2 After a period of use, the bore surface will wear down, leaving a ridge near the top of the bore which marks the upper limit of travel of the top piston ring. The wear ridge may not be readily visible, but it can usually be felt if a fingernail is run along the bore surface. Accurate bore measurement requires the use of a bore micrometer. The measurement should be made at three points along the length of the bore; the top; middle and bottom. At each point, two measurements, one at right angles to the other, should be taken. Subtract the smallest reading from the largest to obtain an indication of the extent of the wear.
3 In the absence of a bore micrometer, a rough indication can be obtained by inserting the piston, less rings, into the bore. Measure the gap between the skirt and the bore surface with feeler gauges at the points described above, then subtract the smallest feeler gauge thickness from the largest to obtain the wear figure. Note that this method is far from accurate, since it does not allow for the curved surfaces of the two components being measured. It is recommended that the measurement is checked to confirm the need for a rebore.
4 If the original bore is to be re-used along with a new piston or piston rings, the surface should be honed to remove the shiny, glazed surface. More importantly, the wear ridge at the top of the bore must be removed or the new top ring may break when it strikes it. This is best entrusted to a professional, who will have the necessary de-ridging and honing facilities. In an emergency, the honing operation may be approximated by roughening the bore with a DIY-type glaze-busting tool.
5 Finally, check the cylinder to cylinder head gasket face using a straight edge and feeler gauges, particularly if there is reason to suspect a leak between the two parts. If there is more than 0.05 mm (0.002 in) warpage, the surface must be lapped until it is flat. This can be carried out using a sheet of abrasive paper taped onto a sheet of plate glass. Place the gasket surface on the abrasive and move it in a semi-rotary fashion for a few moments. Examine the surface again, noting any high spots, which will not have been marked by the abrasive. Continue lapping until all high spots are removed. When work on the barrel is complete, wash it thoroughly in petrol or a similar solvent to remove any residual dirt or abrasive. Allow it to dry, then oil the bore surface to prevent rusting and cover with a rag until it is needed during reassembly.

16 Examination and renovation: piston and rings

1 If inspection of the cylinder barrel has revealed damage or wear beyond the specified service limit, it is unlikely that the existing piston will be serviceable, but if bore condition is acceptable, check the piston and rings as described below. It goes without saying that if the piston is obviously badly worn or damaged it will have to be renewed, even if the cylinder bore is serviceable.
2 Start by removing the piston rings. This can be done by very carefully spreading the ends of each ring in turn and easing it off the piston. Whilst this can be done manually, it is safer and often easier to arrange several thin metal strips as shown in the accompanying line drawing and sliding the rings over them. This method is particularly useful where the rings have become gummed in their grooves by accumulated carbon. Old feeler gauge blades are ideal for this purpose.
3 As each ring is removed, note the etched size markings near the gap and on the upper edge. On a well used machine these may be very faint, so as a precaution mark each ring with a spirit-based felt pen to indicate the groove to which it belongs and the top face. The rings, if they are to be refitted, must be installed in their original grooves.
4 Over a period of time the piston will wear down, the wear being most obvious near the bottom of the piston skirt and at right-angles to the gudgeon pin. The appearance of the piston surface will vary between lightly polished to seriously worn or scored, according to the age of the machine and its past history. The extent of wear can be measured using a micrometer, the measurement being taken at a point about 10 mm from the bottom of the skirt and at 90° to the gudgeon pin bore. If the piston and/or bore have worn to give a piston to bore clearance in excess of the normal 0.025 – 0.045 mm (0.00098 – 0.0177 in), check the piston diameter. No service limit figures are available, but the standard dimensions will be found in the specifications at the beginning of this Chapter. If a new piston of the appropriate oversize will bring the piston/bore clearance within tolerance, and given that the bore surface is undamaged, a new piston may be fitted. Failing this, have the cylinder bored to the next oversize and fit a new piston and rings.

5 Where partial seizure has occurred, the piston surface may have 'picked up' slightly. Provided that the damage is not too severe and that the piston is otherwise serviceable, the roughness may be dressed off using a fine file or abrasive paper. Take care not to remove any more material than is essential to achieve a smooth surface, and remember that if the renovation work causes an excessive piston/bore clearance, the piston must be renewed in any case.
6 Clean out any accumulated carbon from the piston ring grooves using a suitable scraper. A section of broken piston ring is ideal for this job, but failing that a small electrical screwdriver can be used. Take great care not to enlarge the grooves during cleaning. The ring to groove clearance of the top and 2nd rings should be measured using feeler gauges and compared with the specified clearance. Fitting new rings to a used piston may resolve the problem of excessive clearance, but only if the ring grooves are relatively unworn. If in doubt, seek the advice of a Yamaha dealer.
7 The rings themselves should present an even, shiny surface on their working faces, without any signs of scores or discoloration. Any discoloured patches indicate leakage of gas past the rings and thus the need for renewal. Ring wear is assessed by placing each ring in turn into the cylinder bore and measuring the ring end gap. Use the crown of the piston to ensure that the ring sits square in the bore, and position it 20 mm (0.8 in) from the top of the bore. If the end gap of any ring exceeds those given in the specifications, renew the rings as a set.
8 If new rings are to be fitted into a used bore, there is a risk of ring breakage if the top ring contacts the wear ridge. The ridge can be removed by using a de-ridging tool, or by honing, and this should preferably be left to a specialist. It will also be necessary to remove the smooth, glazed surface of the bore to enable the new rings to bed in, and this 'glaze busting' operation should be carried out together with any de-ridging required. This work can be carried out by any competent engineering firm, and most Yamaha dealers will be able to arrange this service if required. Do bear in mind that if the honing work is likely to take the bore past its service limit it will be necessary to renew the barrel and piston anyway, so consider this before having the job done.
9 Examine the gudgeon pin and its bore for signs of wear. Service limit figures for the outside diameter of the pin, inside diameter of the gudgeon pin bore, and the clearance between the two are given in the Specifications section. To check these measurements a micrometer and a bore micrometer will be required, and few owners will possess such equipment. In practice there should be little or no discernible free play between the piston and gudgeon pin. If movement can be felt, consult an authorized Yamaha dealer for confirmation.
10 Fit the gudgeon pin into the small-end eye and feel for clearance between the two. If any movement is noted, try a new gudgeon pin. If this fails to resolve the problem, consult an authorized Yamaha dealer for advice. Note that Yamaha make no provision for reconditioning the small-end eye, short of fitting a new connecting rod.

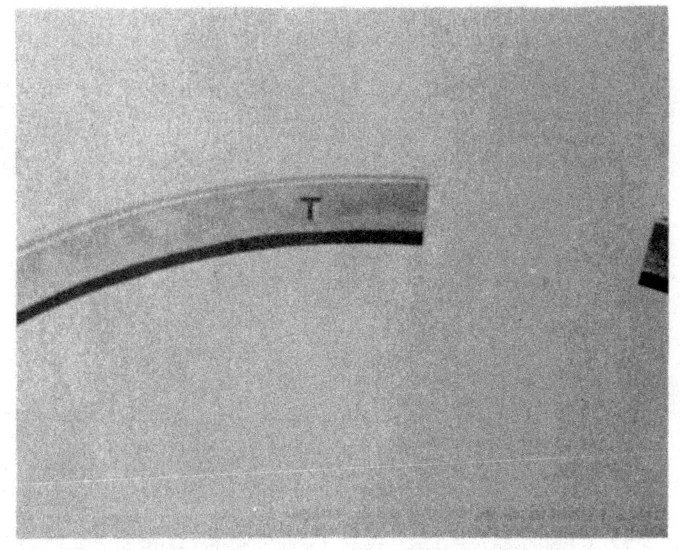

16.3 'T' mark denotes the upper surface of the ring

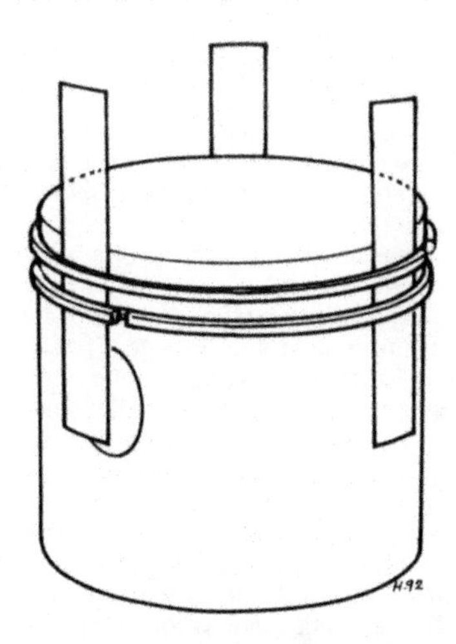
Fig. 1.5 Method of freeing gummed piston rings

17 Examination and renovation: cylinder head and valves

Note: *Many of the operations detailed in this section require specialist equipment and expertise; if you are in any doubt about the procedures described, or lack the necessary equipment, it will be safer and more economical to have the work carried out by a Yamaha dealer.*
1 Having removed the cylinder head assembly as detailed earlier in this Chapter, it must be decided whether or not the valves and/or the camshaft require attention. In the normal course of an overhaul, the cylinder head would be stripped for examination, as it would if there were obvious indications of problems in this area. If, on the other hand, the machine has not covered a great mileage, or if the head, valves and camshaft are known to be in good order, attention can be restricted to cleaning and a cursory visual check. If it is decided that more detailed examination is necessary, proceed as follows. Note that it is possible (although somewhat unconventional!) to remove the camshaft and rockers, but not the valves, with the cylinder head installed.
2 Remove the two bolts which retain the left-hand side cover to the cylinder head and lift it away. Unscrew and remove the two valve inspection caps. The rocker arms pivot on short shafts which are a sliding fit in the head. Yamaha recommend the use of a special tool, in the form of a slide-hammer and attachment to drive the shafts out of the cylinder head. The shafts can also be pulled out of the head casting by screwing an 8 mm bolt into the threaded end of the shaft and pulling them clear. If the shafts prove to be tight in the head, try warming the head casting by immersing it in near boiling water for a few minutes. Once the shafts have been removed, the rocker arms will drop free and can be lifted out.
3 The camshaft is removed in a similar manner, Yamaha recommending the slide-hammer approach. The camshaft can also be withdrawn by warming the head in hot water to loosen the grip on the bearing. If this fails, any slide-hammer can be adapted to fit, using a metric bolt of the correct size screwed into the camshaft end.
4 To remove the valves a valve spring compressor will be required. The valves are quite deeply recessed and of small diameter, so it may be found that many universal compressors will not fit. In the absence of the correct Yamaha tool, try to obtain a compressor designed specifically for motorcycle use.
5 Assemble the compressor, tightening it just enough to compress the springs sufficiently to allow the collet halves to be removed (a small magnet is useful here). If the assembly seems stuck, do not continue tightening the compressor or damage may be caused. Apply reasonable pressure, then tap the end of the compressor to shock the collets out of the spring retainer. Remove the compressor and lift out the retainer, spring, valve stem seal and the lower spring seat. Displace and remove the valve, then place all items in a box to indicate inlet or exhaust, as appropriate. **On no account** mix up the valve components whilst removed. Repeat the removal operation on the remaining valve.
6 Carefully clean all parts prior to inspection. In the case of the cylinder head, remove the accumulated carbon using a scraper and abrasive paper or steel wool, taking care not to score the soft alloy. Wash the decarbonised head and the remaining cylinder head

components in clean petrol and place them on some rag to dry.
7 Examine the cylindeb head closely, looking in particular for signs of cracking around stud holes and the valve seats. If found, seek professional advice at once; renewal of the head may be necessary, in which case it may be worth trying to obtain a serviceable second-hand unit before proceeding further. If the head looks sound, check the gasket face for warpage as described above for the cylinder barrel.
8 Examine the valves in conjunction with the valve seats for wear or damage. If the valves are badly worn, scored or burnt, renewal will be necessary. If the valve seat surface is similarly damaged it may be possible to re-cut it to remove the damage. This is a skilled task best entrusted to an authorized Yamaha dealer who will have the necessary equipment to carry out the work. Note that the valve guides should be checked, and if necessary renewed, **before** the seats are re-cut.
9 The valve stems and guides should be checked for wear, preferably by measurement, and the resulting clearance calculated. In view of the specialist equipment the measurement requires, it may be preferable to have this done by a Yamaha dealer. Failing this, a rough assessment can be made by fitting each valve in its guide and feeling for movement between them. If the guides are worn, renewal, reaming and the necessary seat recutting will need professional attention.
10 Start the checking process with the valves; if they are in need of renewal, Yamaha recommend that the guides should be renewed at the same time. Check the valve stem runout with the valve mounted between V-blocks, using a DTI (dial gauge). The maximum allowable runout figure is 0.02 mm (0.000787 in). The valve, guide and oil seal must also be renewed if the end of the valve has become distorted to a larger diameter than the remainder of the stem.
11 The remaining valve service limits are depicted in the accompanying line drawing; renew the valves and guides as a set if the appropriate valve service limit figure is exceeded in any one dimension. Measure the valve stem diameter at various points using a micrometer, noting the minimum reading obtained. Valve and stem diameters and their respective service limits are given in the Specifications at the beginning of this Chapter. Note that in addition to the individual limits, the clearance between the two must not exceed the service limit.
12 To measure the valve guides, a bore micrometer is required. Once again, compare the largest reading obtained with the maximum figure given in the specifications, and renew the guides as required. Renewal is best entrusted to a Yamaha dealer unless the necessary shouldered drifts, heating facilities and a 5 mm reamer are available. Removal is carried out after warming the head to 100°C (boiling water will suffice for this). Drive out the old guide, then fit a new guide and O-ring. Be absolutely sure that the guide is driven in squarely or the head may be damaged. once fitted, ream the guide bore to 5 mm. After reaming, the valve seat must be re-cut and the valve lapped to the new seat face.
13 The valves and seats should be lapped to ensure a good gastight seal, though it should be noted that any recutting or refacing work must be carried out first. The lapping, or grinding operation will not remove deep pitting. Any attempt to do so will usually result in pocketing of the valve seats, and poor gas flow through the valve.
14 The lapping operation can be carried out manually using a valve grinding tool; a short length of dowel or plastic tube with rubber suckers at each end, and some valve grinding paste. Both can be obtained from car accessory shops at little cost. Lubricate the valve stem with engine oil and introduce it into the guide. Apply a thin film of fine grinding paste to the valve face, then slide the valve home. Fit the grinding tool to the valve head (note that it will not grip the valve unless it has been cleaned properly), then commence grinding.
15 Hold the tool between the palms of the hand and spin it to and fro several times. The tool and valve should then be lifted slightly to redistribute the paste, and the operation repeated. After a few seconds of the lapping operation, remove the valve and wipe it and the valve seat clean. A matt grey line will indicate the contact point between the two. This should be at the mid point of the 45° face of the valve and seat; if it is not the seats should be recut and the valves refaced before proceeding further. The lapping operation should be continued until a thin, unbroken grey line can be seen running around both surfaces. When complete, make sure that every trace of the paste is removed from both valves and seats; failure to do so will allow abrasive to be washed into the lubrication system with disastrous consequences.
16 The valve springs should each be measured to assess their free length. This gives a valuable indication of the condition of the springs; if any have compressed to or beyond the service limit specified, they should be renewed as a set. The valve stem oil seals should be renewed as a matter of course during reassembly.
17 Examine the internal bore and the cam lobe contact surface of the rocker arms, looking for signs of scoring or severe wear. If damaged in any way, renew the rocker arm(s). Check the inside diameter of each rocker arm and the corresponding outside diameter of each rocker shaft, using internal and external micrometers. If either exceeds the service limit shown in the specifications, renew the damaged part. Check that the combined clearance between each rocker and its shaft is within limits.
18 Examine the cam lobes for signs of wear or damage, particularly if this has been found on the rockers. The lobe faces should be smooth and polished; any blueing, pitting or scoring being indicative of the need for renewal. Damage is normally quite obvious, with the tips of the lobes appearing mis-shapen and having a grainy appearance where the lobes have worn through their hardened outer surface. The camshaft must be renewed if worn in this way, or if the overall lobe height is at or lower than the service limits given in the specifications. The degree of normal wear can be assessed by measuring the lobe heights and base circle diameter of each lobe, using a micrometer. Compare the readings obtained with those given in the Specifications, and renew the camshaft if worn beyond the service limit. Clean the camshaft, and ensure that the internal oilways are clear.
19 The cam chain does not normally wear significantly in service, and no wear limit figures are available. If, however, it has been difficult to obtain correct cam chain adjustment, it is probable that the chain is in need of renewal. Where other engine damage indicates oil starvation or seizure at some time, it is prudent to renew the chain as a precaution. The same applies to the cam sprockets; renew them if obviously worn or damaged, and whenever a new chain is installed. The crankshaft-mounted drive sprocket is covered in 14.
20 Check the cam chain tensioner and guide blades, renewing them if badly worn or deeply grooved. Note that if they become seriously damaged, accurate cam chain tension will be impossible to achieve. Inspect the tensioner plunger, spring and its lockscrew and nuts. Renew any damaged or excessively worn part.
21 After examination and renovation of the cylinder head has been completed, the valves, rockers and camshaft can be refitted. Lubricate and then install the valves, using new stem seals. Refit the lower spring seats, the springs (closer wound coils towards the valve head) and retainers, then compress the springs to allow the collet halves to be refitted. Check that these are seated securely as the compressor is removed. To check this, place the head on wooden blocks so that the valve heads are well clear of the workbench, then tap each stem in turn with a plastic-faced hammer.
22 Coat the cam lobes and bearing surfaces with molybdenum disulphide grease, then slide the cam into position. Lubricate the rocker arms and their shafts with engine oil, then refit them in their correct positions. Refit the left-hand cover to the cylinder head, using a new O-ring seal, and tighten the two retaining bolts evenly.

17.2 Use bolt screwed into threaded end of rocker shaft to pull it out of the cylinder head

17.3 Warm cylinder head to facilitate removal of the camshaft

17.5a Assemble compressor, tightening it until the collet halves can be removed ...

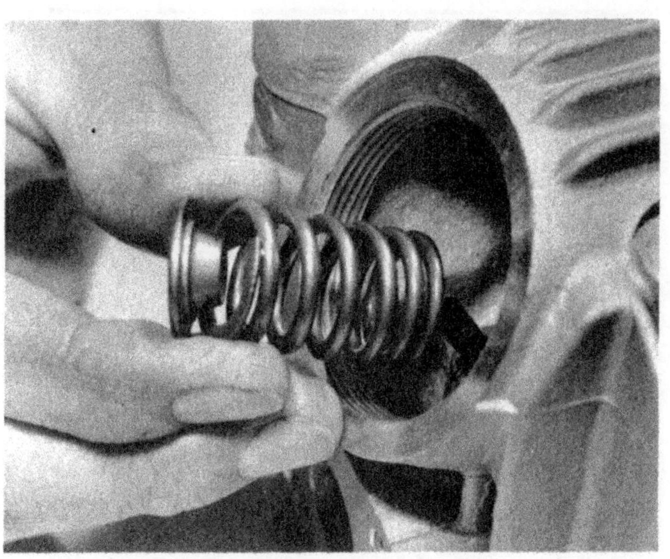

17.5b ... followed by the retainer and spring

17.5c Displace and remove the valve ...

17.5d ... then remove the valve stem oil seal from the top of the guide ...

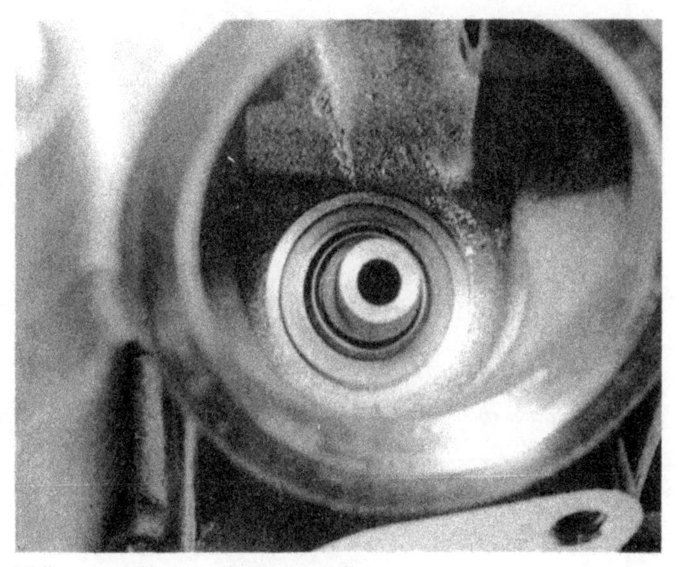

17.5e ... and lift away the lower spring seat

17.18 Check the cam lobe surfaces and cam bearings for wear

17.22 Offer up the rocker arms and refit their shafts

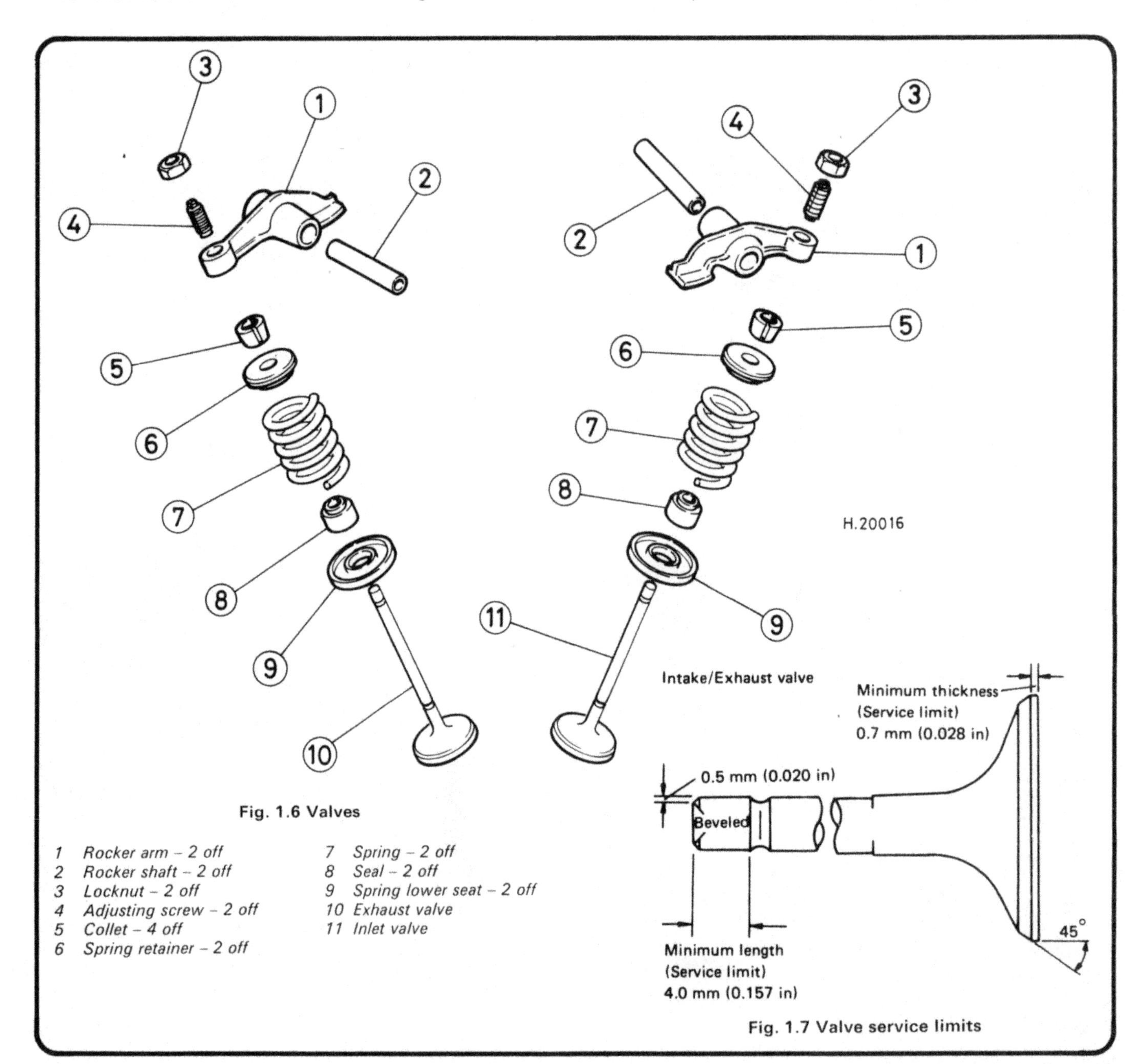

Fig. 1.6 Valves

1 *Rocker arm – 2 off*
2 *Rocker shaft – 2 off*
3 *Locknut – 2 off*
4 *Adjusting screw – 2 off*
5 *Collet – 4 off*
6 *Spring retainer – 2 off*
7 *Spring – 2 off*
8 *Seal – 2 off*
9 *Spring lower seat – 2 off*
10 *Exhaust valve*
11 *Inlet valve*

Fig. 1.7 Valve service limits

18 Examination and renovation: centrifugal clutch, primary drive and centrifugal filter

1 As has been mentioned, the clutch assembly operates automatically, coming into engagement as the engine speed rises and the steel balls contained in the housing are thrown out under centrifugal force. An element of rider control is provided by the linkage connecting the clutch assembly to the gearchange pedal. This disengages the clutch during each gearchange. The clutch is rather more complex than a manual clutch, but poses no special problems from the point of view of overhaul.

Clutch outer drum

2 Examine the drum slots for signs of wear or damage. Indentations in the edges of the tangs with which the clutch plates engage can cause erratic operation and must be rectified. If slight, any damage can be dressed out with a fine file, but more severe wear or damage will necessitate renewal of the drum.

3 Check the fit between the phosphor-bronze bush in the centre of the clutch drum and the sleeve on which it runs. No service limit figures are available, but renew the bush and/or the sleeve if it is excessively loose. If in doubt, check with a Yamaha dealer. A caged roller thrust bearing resides in a recess at the bottom of the drum. It is sandwiched between two plain washers, the assembly being held in position by a round wire circlip. Prise out the circlip, then remove and check the condition of the bearing and washers. Renew them as a set if worn or damaged.

Clutch centre

4 Examine the splines of the clutch centre for wear or damage. As with the drum, light damage may be dressed out with a fine file, but more serious damage will require renewal. Check also the cam slots in the clutch centre, renewing it if these are badly worn or distorted.

One-way boss

5 The one-way boss fits against the clutch centre, its cams engaging in the slots mentioned above. Check the cams for damage, and the fit between the two parts, renewing either or both parts if obviously worn.

Clutch plain and friction plates

6 Check the plates for signs of wear, noting any indication of overheating, such as blueing of the metal components. Check the plain plates for warpage using a surface plate and feeler gauges. Renew the plain plates as a set if any exceed the maximum warpage limit of 0.06 mm (0.0024 in).

7 The friction plates should be checked for wear using a vernier caliper to measure the friction material thickness. Renew the plates as a set if any are worn to 2.9 mm (0.114 in) or less.

Clutch pressure plate assembly

8 The pressure plate assembly is of sealed construction, and apart from checking the bearing, little need be done during overhaul. Needless to say, if the assembly is obviously damaged in some way, it should be renewed. The release bearing is carried in a cast alloy bearing block attached to the pressure plate by four screws. Check that the bearing turns smoothly and easily, with no signs of grittiness. If the bearing is worn, remove the bearing block from the pressure plate to gain access to the bearing. Heat the bearing block in near boiling water, then tap out the old bearing. Reverse this process to fit the new bearing, ensuring that it enters the bore squarely.

Clutch pushrods

9 Check the long and short pushrods for wear, paying particular attention to their ends. Wear is most likely to show up as blueing of the ends, due to heat build-up caused by friction. This can often occur if the release mechanism is set up incorrectly, leaving insufficient clearance between the pushrods. Check the long pushrod for bending by rolling it on a flat surface. If any bending exceeds 0.5 mm (0.02 in) renew the pushrod.

Clutch release mechanism

10 Examine each of the centrifugal release balls for pitting or other damage, renewing them as a set if any are marked in this way. Check the condition of the external release components, looking for stiff or erratic operation or obvious signs of wear. Renew as required.

Clutch springs

11 Measure the free length of each of the clutch springs, renewing them as a set if any one has compressed to 29.1 mm (1.15 in) or less.

Primary drive

12 Check the condition of the primary drive gear teeth, in conjunction with those around the clutch drum. If there are signs of pitting or chipped teeth, renew the clutch drum and primary drive gear as a pair.

Centrifugal oil filter

13 The pressed steel lid of the centrifugal filter should be prised off and any accumulated sludge cleaned out. Note that any metal particles which may have resulted from some component failure will tend to be trapped in this filter, and such debris may be of assistance when attempting to diagnose the cause of the problem. Once the filter is clean, refit the lid and place it to one side until required during reassembly.

18.13 Prise off the centrifugal oil filter cover and remove any accumulated deposits

19 Examination and renovation: gearbox components

Selector drum, forks and shaft

1 Check the selector forks and shaft for wear, ensuring that each fork slides smoothly on the shaft but without excessive free play. Check that the shaft is straight by rolling it on a flat surface. Renew it if it has become bent. The pin on each fork should be checked for wear; it should be a good fit in its groove in the selector drum. If wear has occurred, it will be evident in the form of flats worn on the selector groove pins. This problem can be resolved by renewing the forks, provided that the selector fork grooves are relatively unworn. In extreme cases it may be necessary to renew the drum together with the forks.

Selector mechanism

2 The selector mechanism is unlikely to suffer rapid wear, but if the claw spring or centring spring are weak or broken, they must be renewed. The claw ends should be checked for burring and wear. If they have become rounded off, gear selection problems will have been evident, with gears failing to engage or not engaging fully. It may be possible to restore their profile by careful filing. Failing this, the selector shaft assembly should be renewed.

Input and output shafts – general

3 The gearbox shafts should be checked for trueness by spinning them between centres and measuring any runout with a dial gauge. If runout of either shaft exceeds 0.08 mm (0.0031 in), the shaft must be renewed. Examine the teeth of each gear in turn, looking for chipped teeth, or teeth which have worn through the case hardening. This is unlikely to occur unless the engine has been run with little or no oil at some stage. If damaged teeth are discovered, renew the affected input and output shaft gears as a pair.

4 Check that each gear moves smoothly on its shaft without excessive free play. Again, this is unlikely to be a problem unless the engine has run out of oil. Check the condition of the output shaft bevel gear in conjunction with its middle gear counterpart. If wear is discovered, renew the gears as a set. The gear shafts should not be dismantled unless wear or damage necessitates this.

Input and output shaft dismantling – T50 model

5 The input shaft is a pressed up assembly, for which individual components are not available as spare parts. If damage is apparent the complete input shaft assembly must be renewed.

6 To dismantle the output shaft, remove the circlip from its left-hand end and withdraw the washer and 1st gear pinion, followed by the 3rd gear pinion. Remove the second circlip and splined washer to free the 2nd gear pinion and the bearing plate. The output bevel gear is integral with the shaft and cannot be removed separately.

Input and output shaft dismantling – T80 models

7 In the case of the input shaft, remove the circlip and plain washer from the left-hand end of the shaft assembly, then slide off the 2nd gear pinion. This is followed by the 3rd gear pinion, a second circlip and the 4th gear pinion. The 1st gear pinion is integral with the shaft and thus must be renewed as a unit if worn or damaged.

8 Moving to the output shaft, remove the circlip from its right-hand end, then lift away the plain washer and the double 1st gear and kickstart driven gear. This is followed by a second plain washer and the 4th gear pinion. Work off the remaining circlip and slide off the 3rd gear pinion. The 2nd gear pinion, output shaft, bearing plate and output bevel gear form a pressed-up assembly. It is not possible to obtain individual parts of this assembly, so if any part is worn, the complete unit must be renewed.

Input and output shaft reassembly – all models

9 Reassembly should be carried out using new circlips. Follow the dismantling sequence in reverse, referring to the accompanying line drawing and photographic sequence (T80 only) for details. When fitting circlips do not open them any more than necessary to slide them over the shaft, and make sure that they seat fully in their grooves. Lubricate each gear and the shafts with engine oil as assembly progresses. Check that the gears move freely on the assembled shafts.

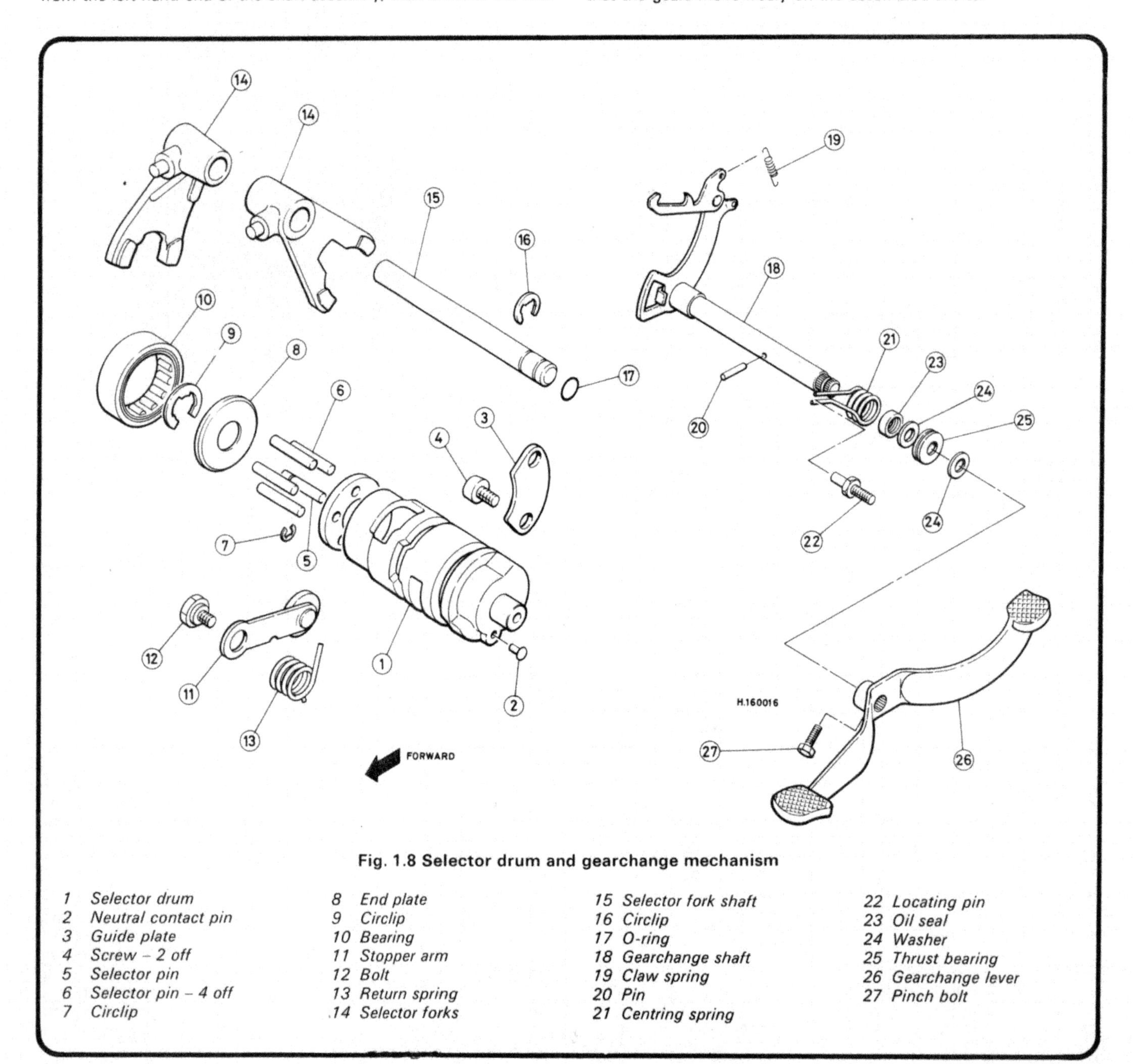

Fig. 1.8 Selector drum and gearchange mechanism

1 Selector drum
2 Neutral contact pin
3 Guide plate
4 Screw – 2 off
5 Selector pin
6 Selector pin – 4 off
7 Circlip
8 End plate
9 Circlip
10 Bearing
11 Stopper arm
12 Bolt
13 Return spring
14 Selector forks
15 Selector fork shaft
16 Circlip
17 O-ring
18 Gearchange shaft
19 Claw spring
20 Pin
21 Centring spring
22 Locating pin
23 Oil seal
24 Washer
25 Thrust bearing
26 Gearchange lever
27 Pinch bolt

19.9a Fit the input shaft 4th gear pinion, retaining it with its circlip (T80)

19.9b Slide on the 3rd gear pinion, selector groove and dogs innermost (T80)

19.9c Refit the 2nd gear pinion (T80) ...

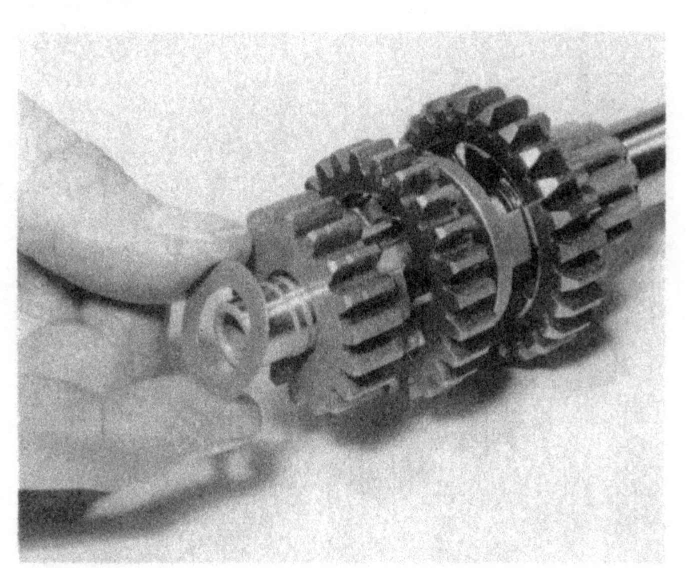

19.9d ... followed by the large plain thrust washer (T80) ...

19.9e ... and secure the assembly with a circlip (T80)

19.9f The 2nd gear pinion forms an assembly together with the output shaft, bearing and retainer plate (T80)

19.9g Slide the 3rd gear pinion over the shaft as shown (T80) ...

19.9h ... and secure it with its circlip (T80)

19.9i Fit the 4th gear pinion, selector groove innermost (T80)

19.9j Place the plain washer against the shouldered shaft end (T80)

19.9k Fit the combined 1st gear and kickstart driven gear (T80) ...

19.9l ... followed by the plain washer and circlip (T80)

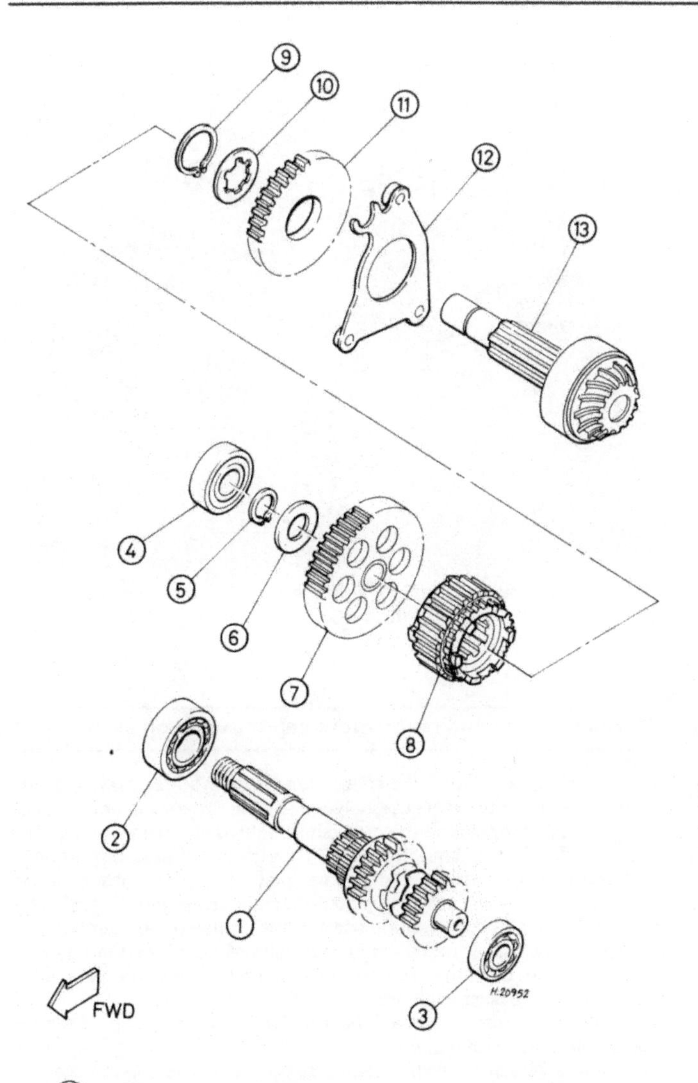

Fig. 1.9 Gearbox shafts – T50 model

1 Input shaft assembly
2 Input shaft right-hand bearing
3 Input shaft left-hand bearing
4 Output shaft right-hand bearing
5 Circlip
6 Washer
7 Output shaft 1st gear pinion
8 Output shaft 3rd gear pinion
9 Circlip
10 Splined washer
11 Output shaft 2nd gear pinion
12 Bearing plate
13 Output shaft, bearing and bevel gear

20 Examination and renovation: kickstart mechanism

1 Examine the teeth of the large kickstart pinion and the driven pinion on the transmission output shaft for wear or damage. Problems such as the mechanism slipping or jamming can often be attributed to worn or chipped teeth, and if such damage is discovered, renew the gears as a pair.

2 Check the ratchet teeth on the kickstart pinion and the ratchet block for wear or damage. If the teeth are rounded off or worn down, slipping is likely to occur and both items should be renewed. Check the fit of the ratchet block in relation to the splines on the kickstart shaft. Wear is unlikely here, but if it does occur the assembly may be inclined to jam occasionally.

3 A broken return spring will be self-evident, and this item can be renewed without the need for crankcase separation, should the need arise. Check also that the friction clip is a good fit on its boss. This can be checked by measuring with a spring balance the pressure required to move the clip. This should be 0.8 – 1.5 kg (1.76 – 3.31 lb). If too tight or too loose, do not attempt to bend the clip to correct the fault; it should be renewed.

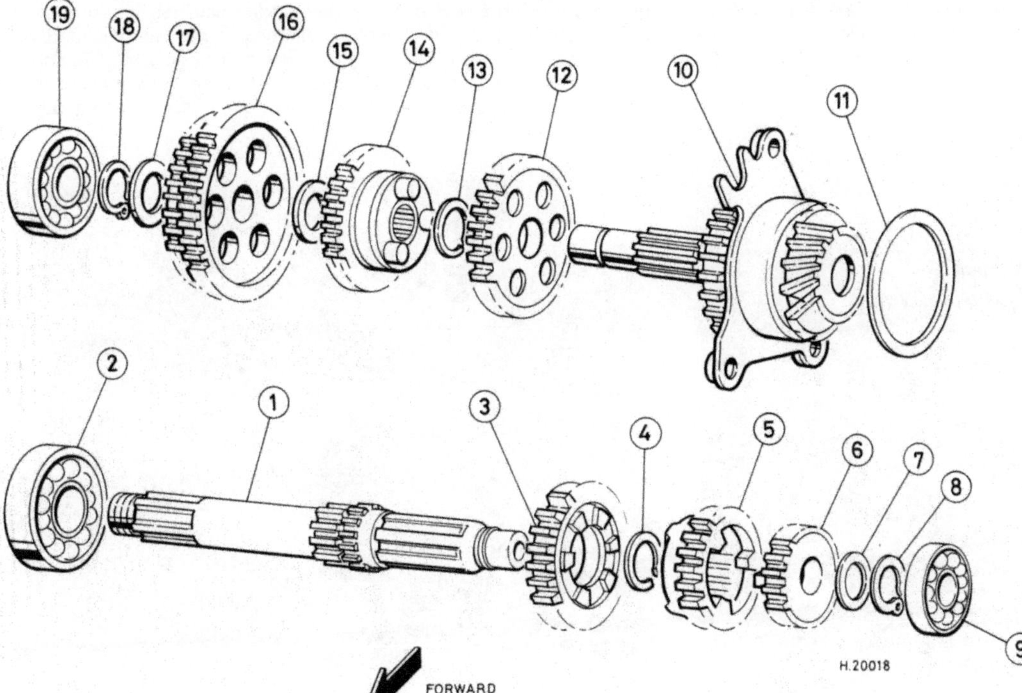

1 Input shaft with 1st gear pinion
2 Right-hand bearing
3 Input shaft 4th gear pinion
4 Circlip
5 Input shaft 3rd gear pinion
6 Inut shaft 2nd gear pinion
7 Thrust washer
8 Circlip
9 Left-hand bearing
10 Output shaft with 2nd gear pinion, bearing and retainer
11 Shim
12 Output shaft 3rd gear pinion
13 Circlip
14 Output shaft 4th gear pinion
15 Washer
16 Output shaft 1st gear pinion and kickstart driven gear
17 Washer
18 Circlip
19 Right-hand bearing

Fig. 1.10 Gearbox shafts – T80 model

20.1 Check the condition of the kickstart gear and ratchet teeth. Renew the spring if stretched or distorted

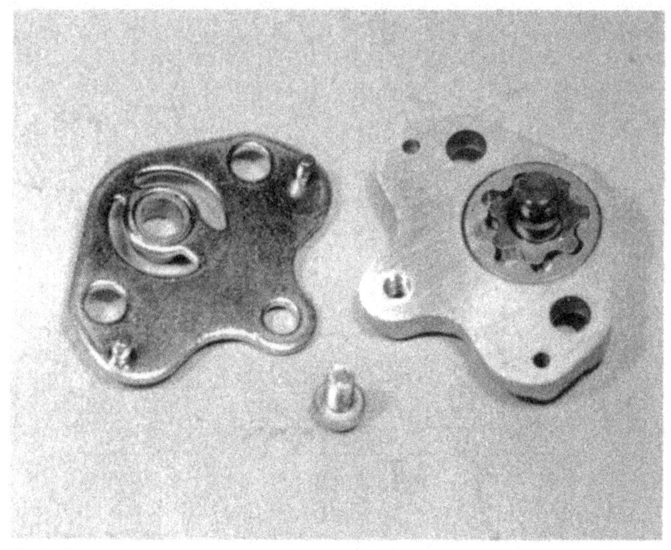

21.2 Remove the oil pump cover and check the rotor and body clearances as described in the text

21 Examination and renovation: oil pump

1 The oil pump should be checked during engine overhauls, and at any time a problem with the lubrication system is suspected. Bear in mind that the engine is completely dependent on its lubrication supply; if it fails serious engine damage is inevitable.
2 With the pump removed from the crankcase as described earlier in this Chapter, remove the single screw which retains the end cover. Lift this away to gain access to the pump rotors. Examine the rotors and the pump body for signs of damage. Scoring, caused by contaminants in the oil, will necessitate renewal of the pump. If the pump is undamaged, check for wear by measuring with feeler gauges the clearance between the outer rotor and the body, the inner rotor and outer rotor (between the rotor tips) and the rotor endfloat. The last check is carried out by placing a straightedge across the pump body and using a feeler gauge to assess wear. Compare the readings obtained with the service limit figures given in the Specifications section of Chapter 2, and renew the pump assembly if in excess of these.

22 Examination and renovation: bearings and oil seals

1 Before reassembly commences, check the various bearings and seals housed in the crankcase halves for wear. These include most of the gearbox bearings and the crankshaft right-hand main bearing. The bearings should turn smoothly, with no sign of stiffness or grittiness, and without discernible radial free play. If in any doubt, seek professional advice or play safe and renew the bearing in question. It is normal practice to renew all oil seals in the course of an overhaul; the old seals may appear undamaged, but will have hardened in use. It is better to renew them than risk seal failure and oil leaks after the engine unit has been reassembled and installed.
2 Once the new seals and any bearings required have been obtained, make a note of the direction of fitting of the oil seals, then lever them out using a screwdriver blade and discard them. The bearings can be driven out using a socket of slightly smaller diameter than that of the bearing outer race as a drift. Note which way each bearing is fitted; one side carries the manufacturer's part number. The new bearing should be refitted facing in the same direction. Make sure that the area around the bearing boss is well supported on wooden blocks to avoid placing

22.1 Examine the various crankcase bearings and seals for wear and renew them as required

22.2a Remove retainers, where appropriate, to permit bearing removal

undue strain on the crankcase castings. To ease removal and refitting, warm the crankcase halves first in very hot water to expand the alloy.
3 Fit the new bearings with the manufacturer's name and part number facing in the correct direction, and make sure that the bearing is tapped home squarely into its bore. Check that it seats fully, and use only the **outer** race when driving the bearings into position. The new bearings should be lubricated with engine oil. The seals should also be fitted facing in the correct direction, and care should be taken not to distort them. Wipe the seal lips with lithium or molybdenum grease.

22.2b Where bearings are fitted in blind bores, a proprietary bearing extractor and slide-hammer should be used for removal

23 Reassembling the engine/transmission unit: general

1 Before reassembly of the engine/transmission unit is commenced, the various component parts should be cleaned thoroughly and placed on a sheet of clean paper, close to the working area.
2 Make sure all traces of old gaskets have been removed and that the mating surfaces are clean and undamaged. Great care should be taken when removing old gasket compound not to damage the mating surface. Most gasket compounds can be softened using a suitable solvent such as methylated spirits, acetone or cellulose thinner. The type of solvent required will depend on the type of compound used. Gasket compound of the non-hardening type can be removed using a soft brass-wire brush of the type used for cleaning suede shoes. A considerable amount of scrubbing can take place without fear of harming the mating surfaces. Some difficulty may be encountered when attempting to remove gaskets of the self-vulcanising type, the use of which is becoming widespread, particularly as cylinder head and base gaskets. The gasket should be pared from the mating surface using a scalpel or a small chisel with a finely honed edge. Do not, however, resort to scraping with a sharp instrument unless necessary.
3 Gather together all the necessary tools and have available an oil can filled with clean engine oil. Make sure that all new gaskets and oil seals are to hand, also all replacement parts required. Nothing is more frustrating than having to stop in the middle of a reassembly sequence because a vital gasket or replacement has been overlooked. As a general rule each moving engine component should be lubricated thoroughly as it is fitted into position.
4 Make sure that the reassembly area is clean and that there is adequate working space. Refer to the torque and clearance setting wherever they are given. Many of the smaller bolts are easily sheared if overtightened. Always use the correct size screwdriver bit for the cross-head screws and never an ordinary screwdriver or punch. If the existing screws show evidence of maltreatment in the past, it is advisable to renew them as a complete set.

24 Reassembling the engine/transmission unit: refitting the crankcase components and joining the crankcase halves

1 Before commencing work, make sure that the crankcase halves, the crankshaft assembly and the gearbox components are clean and grease-free. If any of the crankcase bearings were removed due to wear or damage, fit the new bearings into their respective bores, driving them home squarely using a large socket as a drift against the bearing outer race; see Section 22. Refit any bearing retainers, tightening their retaining screws securely.
2 Arrange the crankcase left-hand casing half on the workbench, supporting it on wooden blocks to allow room for the crankshaft to protrude when fitted. Place the gearbox output shaft shim against its bearing boss. Assemble the gearbox input and output shafts in their correct relative positions on the bench. Fit a new O-ring and the circlip on the end of the selector fork shaft. Slide the selector forks onto the shaft. It is important that the forks are fitted in the correct position. Looking along the shaft from the circlip end, and with the forks hanging down, check that the longer bosses face the end of the shaft and that the locating pegs point to the left (see accompanying photograph).
3 Offer up the selector fork and shaft assembly, making sure that the fork at the circlip end of the shaft engages in the input shaft 3rd gear groove and that the other fork locates over the output shaft 4th gear groove. Make sure that the output shaft bearing plate slot locates over the end of the selector fork shaft. Now position the selector drum, locating it over the selector fork pins, and lower the assembly into the crankcase left-hand half. Check that the various gears engage properly by operating the selector drum by hand. It is a good idea at this stage to retain this assembly by fitting the three bolts which hold the output shaft bearing plate finger-tight (see photograph). Note that each bolt should be fitted with a new copper sealing washer.
4 Lubricate the crankshaft main bearings with clean engine oil, then offer up the crankshaft. If it fails to seat fully, check that the main bearing outer race is entering its recess squarely. Check that the connecting rod is positioned within the crankcase mouth area. Check that all bearings and retainers are in place in the crankcase right-hand half, and apply a thin film of Yamaha Bond No. 4 or equivalent to the joint faces.
5 The crankcase right-hand half can now be lowered into position. Check that the two halves align correctly over the dowels. Refit the crankcase screws, noting that screws 3 and 8 in the tightening sequence shown in the accompanying line drawing should be fitted with copper washers. Any cable or wiring should be refitted in the positions noted during dismantling. Tighten the screws evenly and progressively to pull the cases together squarely, following the tightening sequence shown in the accompanying line drawing. Before proceeding further, check that the crankshaft and transmission shafts can rotate smoothly. If they seem tight, separate the crankcase halves and resolve the problem before continuing with reassembly.

24.2a Place the crankcase left-hand half on the workbench, supporting it on wooden blocks

24.2b Fit the gearbox output shaft shim into its casing recess

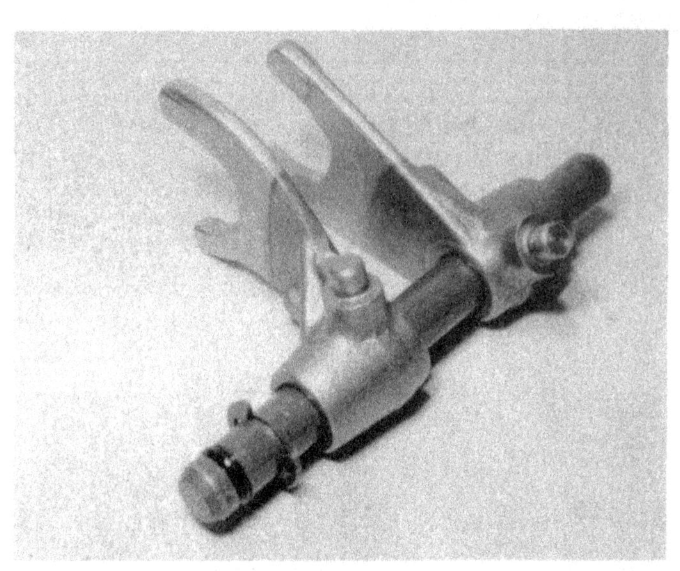
24.2c Assemble the selector forks and shaft in the positions shown

24.3a Assemble the gearbox shafts, selector drum, forks and shaft (T80 shown)

24.3b Make sure that the bearing retainer plate engages the selector fork shaft as shown

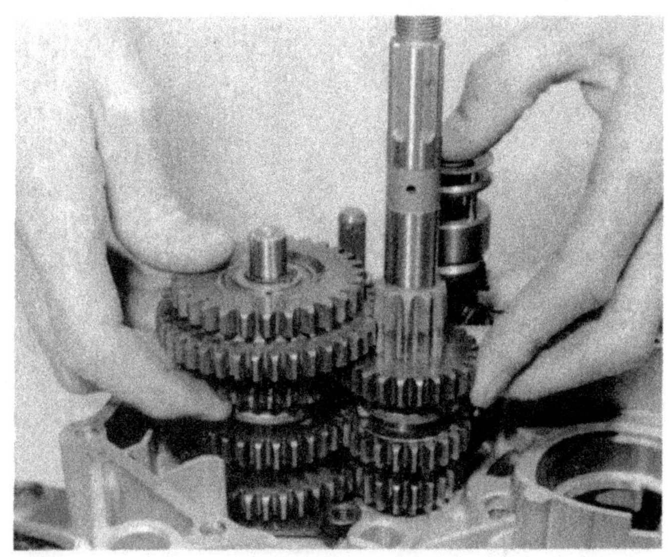
24.3c Place the assembly into the crankcase half, and check that gears operate normally

24.3d Fit the retainer plate bolts, using new copper sealing washers

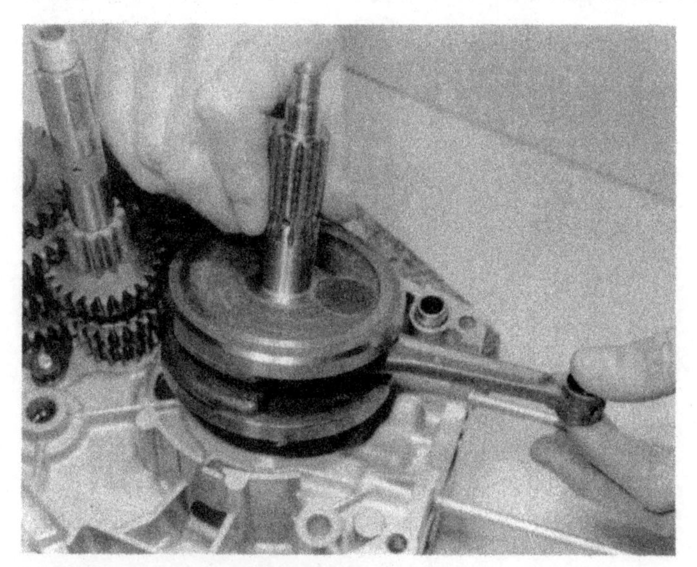

24.4 Offer up the crankshaft assembly, having first lubricated the main and big-end bearings with oil

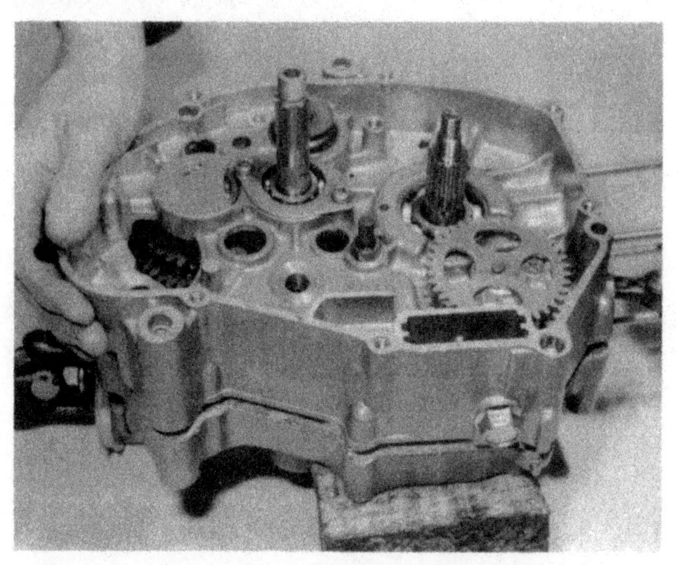

24.5 Join the crankcase halves, then fit and secure the retaining bolts

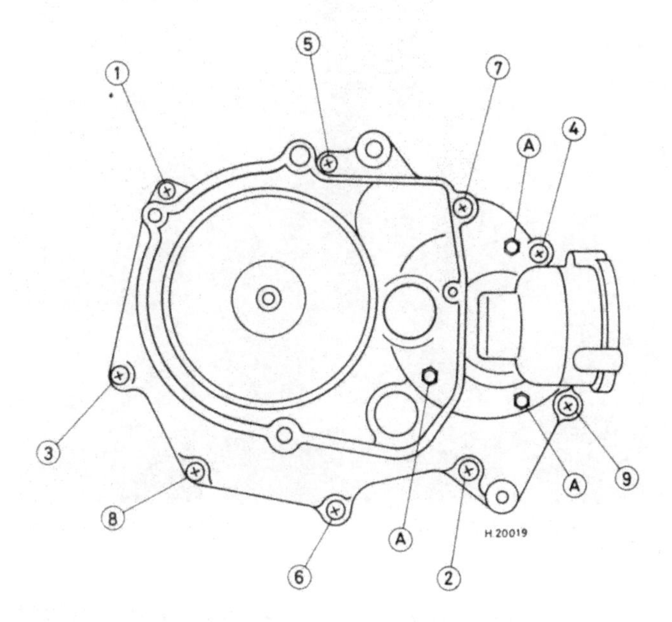

Fig. 1.11 Crankcase screws tightening sequence

Screws 3, 8 and A have copper washers
A – Output shaft bearing plate securing bolts

25 Reassembling the engine/transmission unit: refitting the middle gear assembly

1 If it was removed during the overhaul, the middle gear assembly can be refitted at this stage. It is assumed that the assembly has been left undisturbed. If it was dismantled for any reason it will first be necessary to check the bevel gear teeth mesh depth. This procedure is covered in Chapter 4. **On no account** refit the assembly if it has been dismantled and the mesh depth has yet to be checked.
2 Offer up the assembly, making sure that the bevel gear teeth mesh correctly. Once in place, fit the three bearing plate holding bolts and tighten them evenly and progressively to 1.0 kgf m (7.2 lbf ft).
3 Where appropriate, check that the three bolts which hold the output shaft bearing plate are tightened evenly and securely. Two of these are to be found on either side of the middle gear case projection, whilst the third is situated just above the gearchange shaft bore, refer to Fig. 1.11.

26 Reassembling the engine/transmission unit: refitting the alternator assembly

1 Fit new O-rings to the recesses around the threaded bosses in the crankcase immediately behind the stator; these seal the stator fixing screws. If it was removed, loop the cam chain around the crankshaft end, engaging it over the drive sprocket and feeding the loop of chain out through the crankcase mouth.
2 Offer up the alternator stator and ignition pickup assembly, ensuring that the wiring passes through the casing cutout. Reconnect the CDI and alternator output wiring at their connectors on the CDI and regulator/rectifier units respectively. Refit the neutral switch lead. The stator and pickup are retained as an assembly by two mounting screws. Ensure that the stator seats fully, and that the large O-ring around its edge is in good condition; the stator effectively closes off the entrance into the cam chain drive sprocket area and thus must be oil-tight.
3 Fit the Woodruff key into the slot in the crankshaft end and offer up the alternator rotor. Fit the rotor retaining nut finger-tight, then lock the crankshaft using the holding tool or by passing a bar through the connecting rod small-end eye as described during removal. Tighten the rotor securing nut to 4.0 kgf m (29 lbf ft).

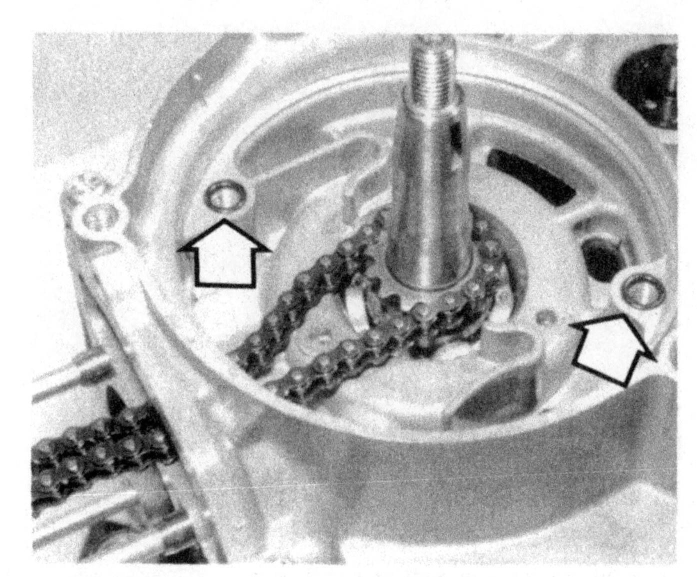

26.1 Fit new O-rings to recesses around screw holes (arrowed). Loop cam chain around crankshaft sprocket

26.2a Check the condition of the O-ring around the alternator stator, renewing it if damaged or stretched

26.2b Offer up the stator, fitting the wiring grommet into its casing slot and then tighten the two retaining screws

26.2c Reconnect the neutral switch lead

26.3a Check that the Woodruff key is in place in the crankshaft end

26.3b Offer up the rotor, and fit the retaining nut ...

26.3c ... tightening it to the prescribed torque setting

27.2a Fit a new oil pump gasket, holding it in place with grease

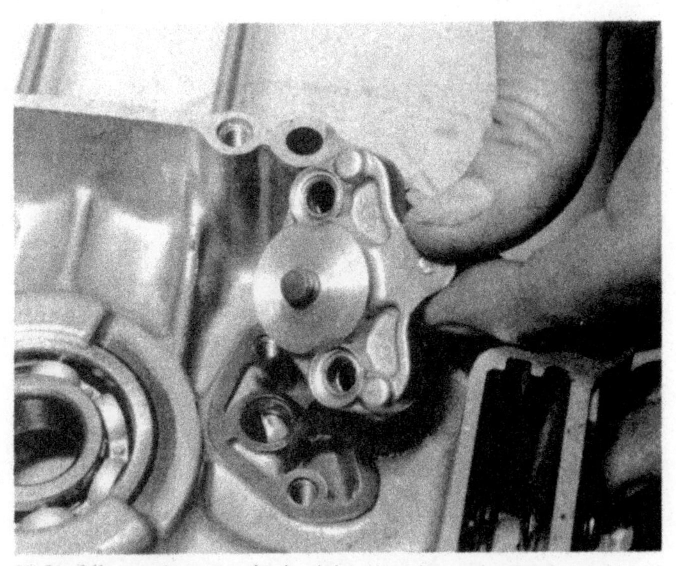
27.2b Offer up the pump body, tightening the retaining screws evenly and securely

27 Reassembling the engine/transmission unit: refitting the oil pump and pickup strainer

1 Check that the oil pump rotors and body are clean, then oil them liberally and assemble the pump. Check that the dowel is in place, then fit the pump cover, securing it with its single retaining screw.
2 Check that the pump gasket faces are clean, then fit a new gasket to the crankcase face, holding it in position with a dab of grease. Offer up the pump assembly and fit the two retaining screws finger-tight. Tighten the screws evenly to 0.7 kgf m (5.1 lbf ft). Offer up the large plain washer, then the pump driven pinion and secure it with its circlip.
3 Clean the oil pickup strainer, making sure that the gauze mesh is intact. Slide the strainer into its slot in the crankcase below the oil pump.

27.2c Place the large plain washer over the pump spindle ...

27.2d ... then fit the pump pinion and circlip

27.3 Fit the oil pickup strainer into its slot at the bottom of the crankcase

28.1a Slide the oil pump drive pinion over the crankshaft splines ...

28.1b ... followed by the centrifugal filter unit, primary drive pinion, washer and retaining nut

28 Reassembling the engine/transmission unit: refitting the centrifugal filter and primary drive gear

If it has not already been done, prise off the pressed steel cover of the centrifugal filter and carefully remove any sludge from the inside of the unit. Once clean, press the cover back into place. Slide the oil pump drive gear over the crankshaft, pushing it down and into mesh with the driven gear. Fit the centrifugal filter, followed by the primary drive gear, washer and retaining nut. Using the same method employed during removal, hold the crankshaft and tighten the retaining nut to 5.0 kgf m (36 lbf ft).

28.1c Lock the crankshaft, and torque tighten the retaining nut

29 Reassembling the engine/transmission unit: refitting the gear selector mechanism

1 Check that all bearing retainers are secure and that their retaining bolts have been tightened. If it was removed, refit the selector drum stopper arm, spring and pivot bolt. The spring end should bear against the crankcase web and the bolt should be tightened to 1.4 kgf m (10 lbf ft).
2 Check that the selector shaft centring spring and claw spring are fitted correctly, then slide the shaft assembly into its casing bore. As it nears the selector drum, lift the claw against spring pressure and into engagement with the drum end. At the same time, make sure that the centring spring fits over its locating pin.
3 Moving to the left-hand side of the unit, place the plain washer, thrust bearing and second plain washer over the selector shaft end, then fit the pin through the shaft to secure them. Offer up the cam, ensuring that it locates over the projecting ends of the pin. Retain the cam with its circlip.

29.1a Refit the selector drum retainer, tightening the mounting screws securely

29.1b Offer up the selector drum stopper arm ...

29.1c ... ensuring that the arm and spring locate as shown

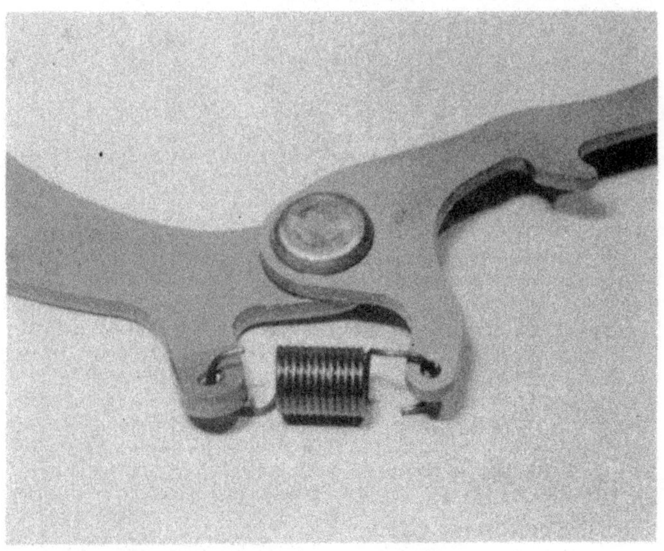
29.2a Check that the selector claw spring is fitted ...

29.2b ... and that the centring spring ends locate as shown

29.2c Offer up the selector shaft assembly, making sure that it engages over the drum end and that the centring spring fits over the locating pin

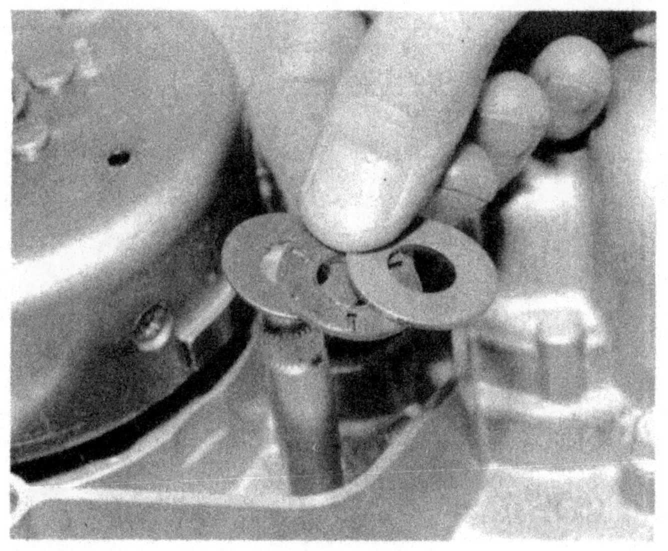
29.3a Place the washers and thrust bearing over the shaft end

29.3b Slide the pin through the shaft hole to retain the assembly

29.3c Place the cam over the shaft, ensuring that it engages over the pin ...

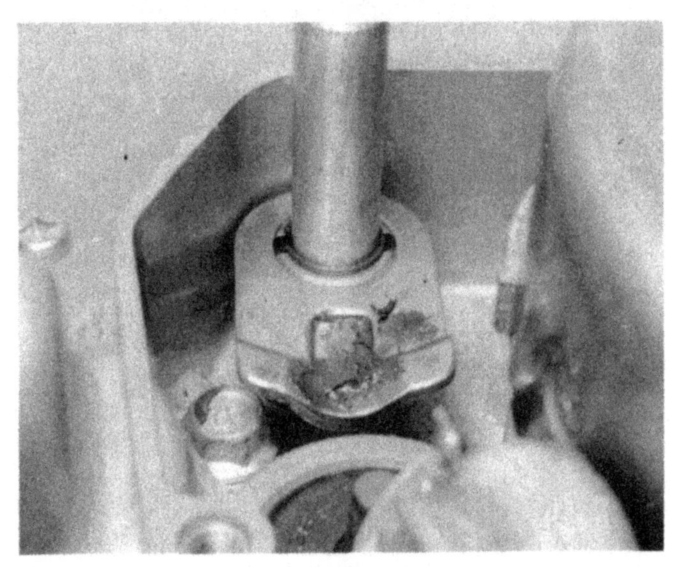

29.3d ... and slide the wire circlip down the shaft to hold the assembly in position

30 Reassembling the engine/transmission unit: refitting the kickstart mechanism

1 If it has been stripped for examination, reassemble the kickstart shaft components as follows. Slide the kickstart pinion, wave washer and plain washer over the shaft, retaining them with the circlip. These are followed by the ratchet block, complete with its friction clip, and retained by a second circlip. Now fit the spring guide, the spring (hooked end first) and the spring cover. Secure these with the third circlip, then fit the final circlip to the remaining groove on the shaft.
2 Offer up the shaft assembly, making sure that the gear teeth mesh and that the spring end is hooked over its projecting pin in the casing. The return spring should be tensioned by turning it clockwise until it can be hooked over the pin. The friction clip must locate in its slot in the crankcase.

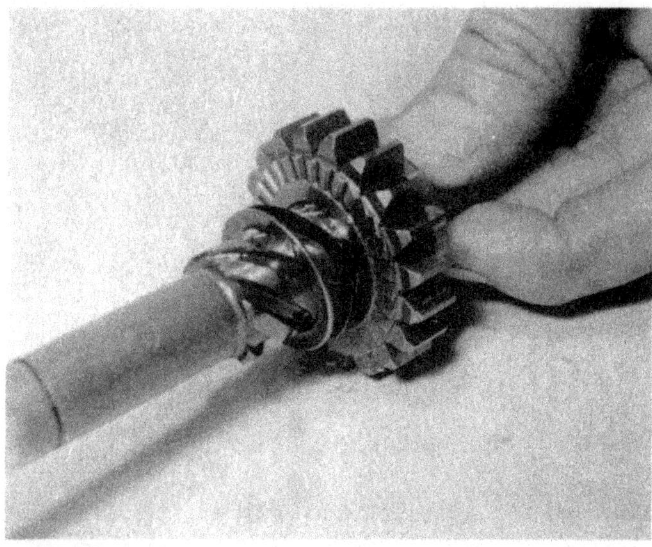

30.1a Fit the kickstart pinion, wave washer and plain washer over the shaft ...

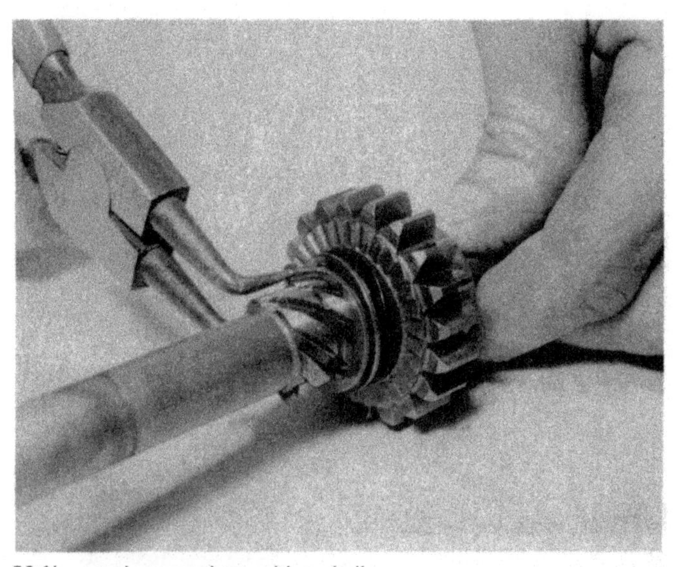

30.1b ... and secure them with a circlip

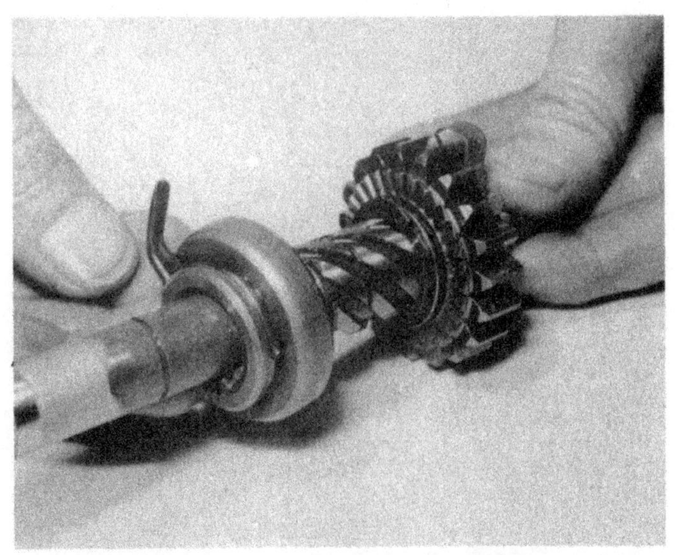

30.1c The ratchet block and its friction clip are fitted next ...

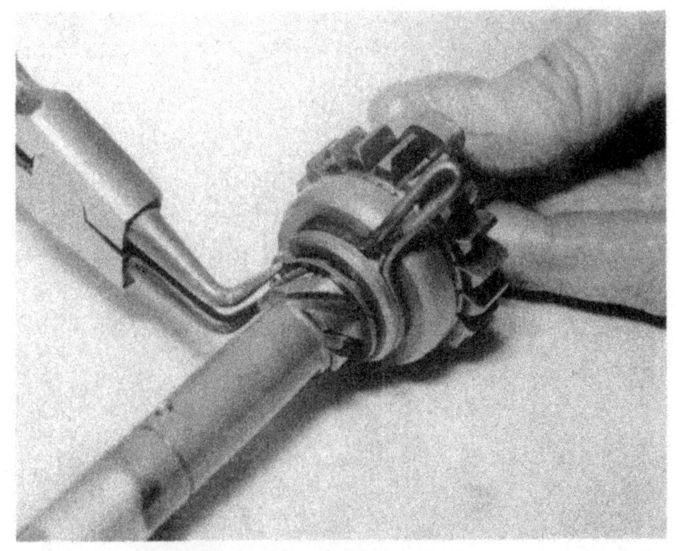

30.1d ... this too being held in place by a circlip

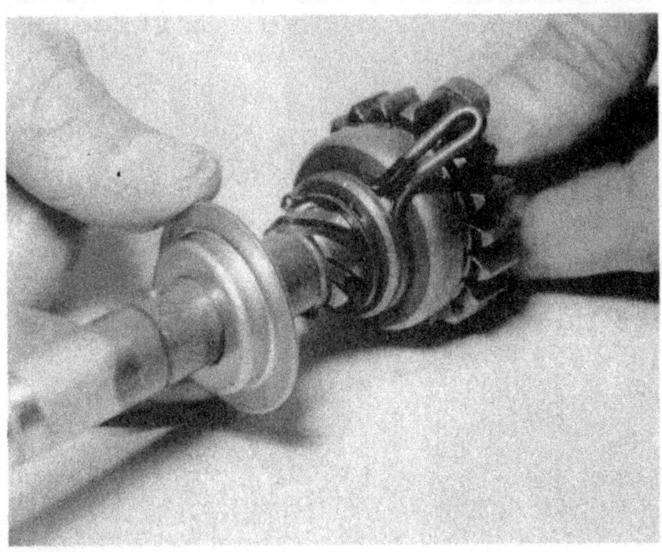

30.1e Now place the spring guide over the shaft ...

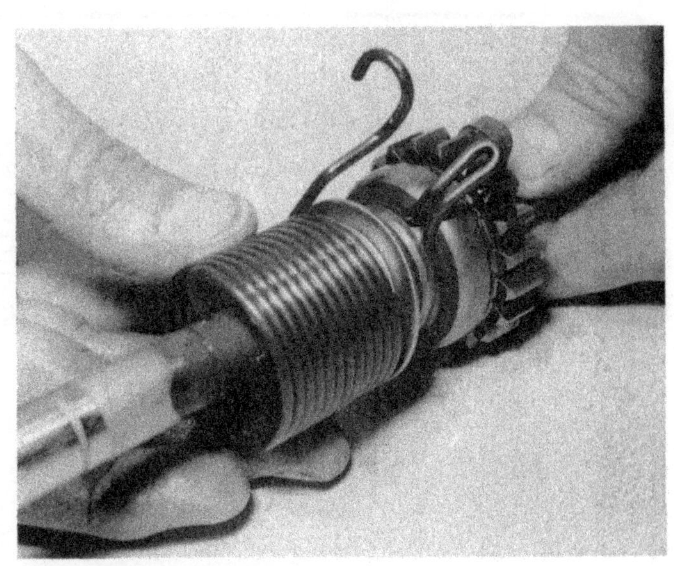

30.1f ... fit the spring, locating the internal tang in the shaft

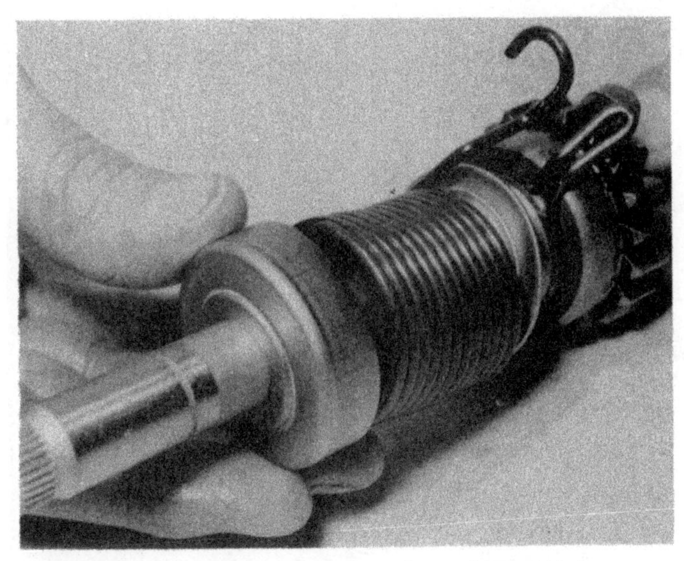

30.1g Place the spring cover over the spring end to hold it in position

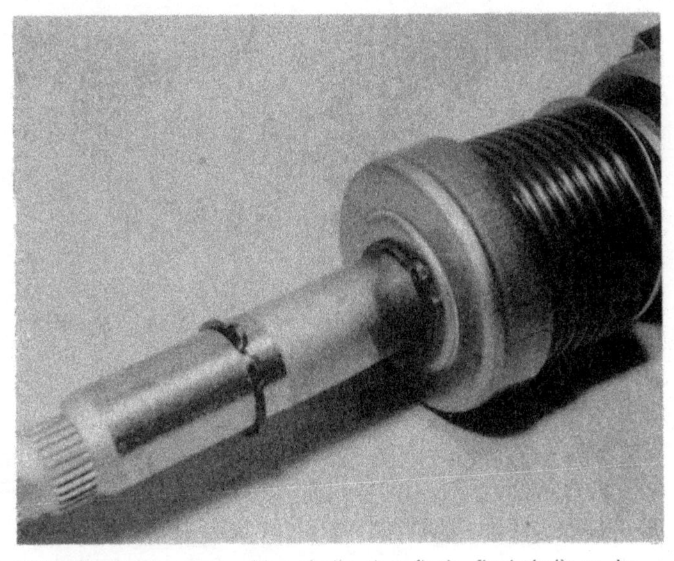

30.1h Retain the cover with a circlip, then fit the final circlip to the remaining shaft groove

30.2a Place the assembly into the casing recess ...

30.2b ... ensuring that the spring end hooks over the pin in the casing and that the friction clip locates properly

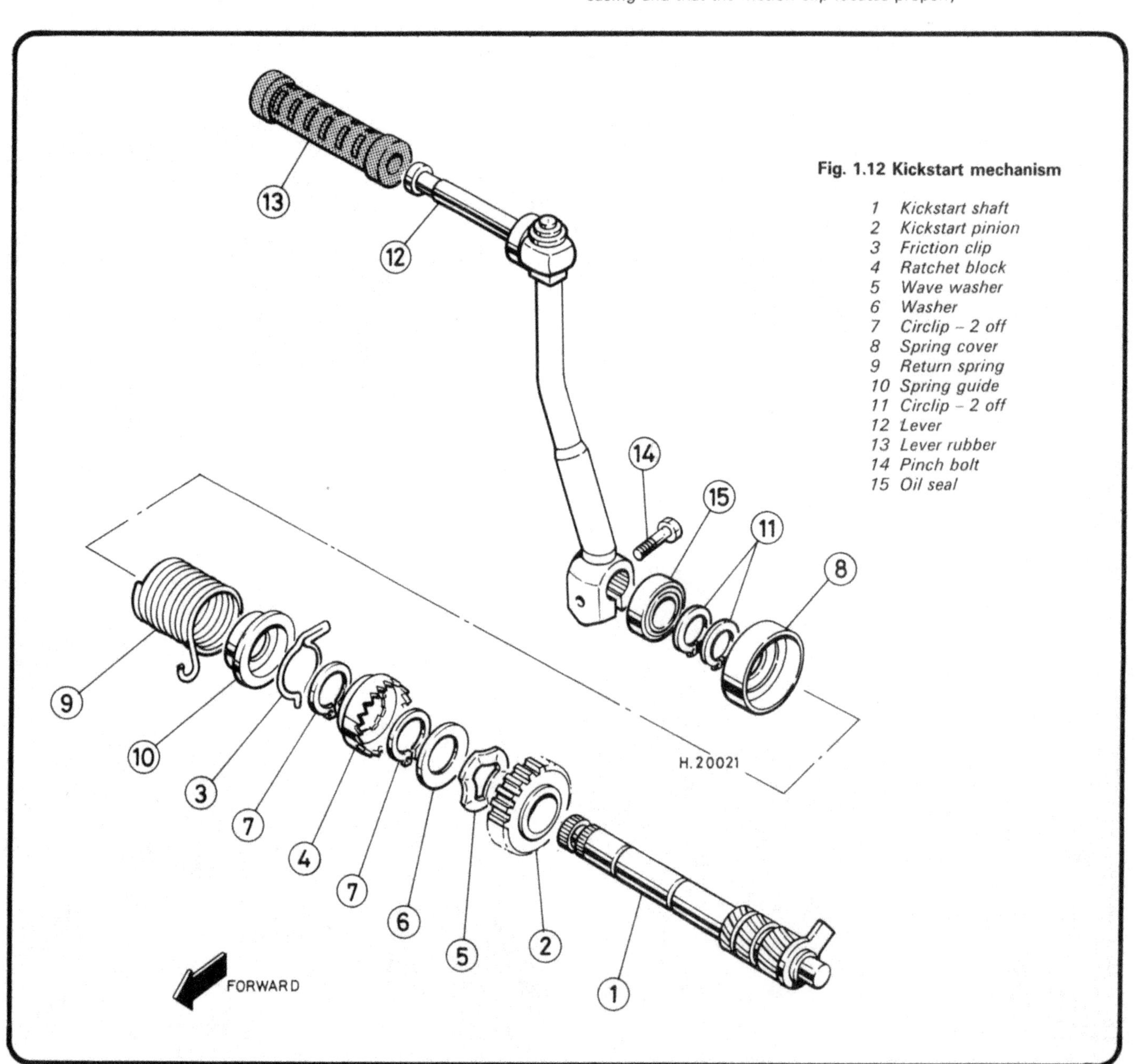

Fig. 1.12 Kickstart mechanism

1 *Kickstart shaft*
2 *Kickstart pinion*
3 *Friction clip*
4 *Ratchet block*
5 *Wave washer*
6 *Washer*
7 *Circlip – 2 off*
8 *Spring cover*
9 *Return spring*
10 *Spring guide*
11 *Circlip – 2 off*
12 *Lever*
13 *Lever rubber*
14 *Pinch bolt*
15 *Oil seal*

31 Reassembling the engine/transmission unit: refitting the centrifugal clutch and crankcase right-hand cover

1 Fit the large plain thrust washer over the end of the gearbox input shaft, followed by the clutch drum sleeve. The outer drum can now be fitted, and the 12 (T50) or 16 (T80) clutch release balls placed in their slots. This last operation will pose problems if the clutch is being rebuilt with the engine unit in the frame. To hold the balls in place, place a dab of grease in each recess.

2 Place the one-way boss over the shaft, followed by the clutch centre. Fit the thrust weight plate over the release balls. It is now necessary to check that the clutch centre is set at the correct height. Hold the thrust weight plate down against the release balls, taking care that it sits flat and does not rock during the check. Using a vernier caliper, measure the difference in height between the upper surface of the thrust weight plate and the flange on the clutch centre. This should be 0.2 – 0.4 mm (0.0078 – 0.0157 in).

3 If the clutch centre height is incorrect, remove the outer drum, and remove the thrust washers, toothed thrust washer – 35T model only and thrust bearing which are retained in its centre by a wire circlip. Measure the thickness of the washers and calculate the required thickness to restore the clutch centre to the correct height. Washers are available in 1.7 mm, 1.9 mm, 2.1 mm and 2.3 mm thicknesses. Obtain the necessary washers, reassemble the clutch drum, release balls, plate and centre, and re-check the centre height.

4 In the middle of the clutch centre, fit the plain washer, wave washer, thrust plate, washer and clutch centre nut. Lock the clutch using the same method employed during removal and tighten the centre nut to 5 kgf m (36 lbf ft).

5 Fit the clutch plates, starting and finishing with a friction plate, and noting that the plain plate with the bent tangs is fitted outermost, and with the tangs angled towards the crankcase. Note the slots in alternate plain plate tangs which should coincide with the spring posts. Check the clutch plate thickness as described below before proceeding further.

6 Once again holding the plates to prevent rocking, use a vernier caliper to measure the difference in height between the top surface of the outer friction plate and the adjacent step in the outside of the clutch drum. If this is outside the range 1.3 – 1.65 mm (0.0512 – 0.0650 in) remove the plates and measure the friction plate thicknesses. The standard plain plate thickness is 1.4 mm, but to permit adjustment, plates are also available in 1.2 mm and 1.6 mm thicknesses. Obtain the necessary plates to set the plate height correctly, then install the plates and re-check the height.

7 Place the clutch springs over the support posts, then fit the long pushrod into the shaft end and the short pushrod into the clutch pressure plate. (If the bearing block was removed from the pressure plate the assembly sequence can be modified as shown in the accompanying photographic sequence.) Offer up the pressure plate assembly and secure it using the large internal circlip. It is easiest to work around the clutch drum in stages, making sure that the circlip seats fully by squeezing it home with pliers (see photograph).

8 Before the outer cover is refitted, check that the oil pickup strainer is in position and that the oilways and seals in the cover are clean and undamaged. Offer up the cover using a new gasket. Fit the cover screws and tighten them evenly to 0.7 kgf m (5.1 lbf ft). Refit the kickstart lever, tightening the pinch bolt to 1.2 kgf m (8.7 lbf ft).

31.1a Place the plain thrust washer over the input shaft end ...

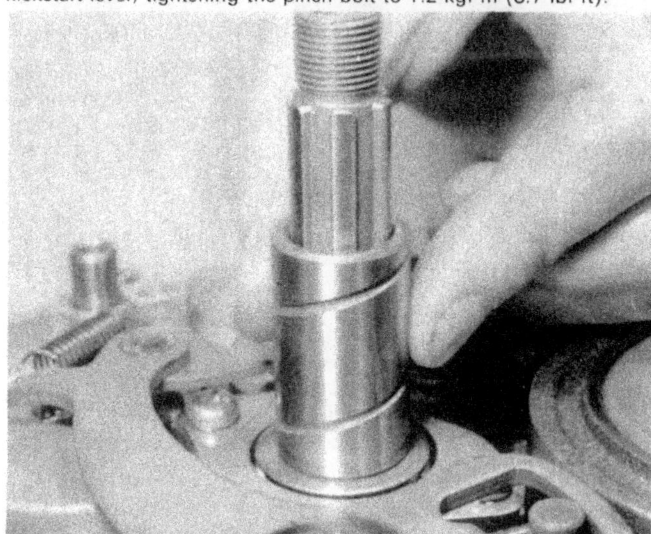

31.1b ... followed by the clutch drum inner sleeve

31.1c Fit the clutch outer drum and install the release balls (T80 shown)

31.2a Place the one-way boss over the shaft end ...

31.2b ... followed by the clutch centre

31.2c Fit the thrust weight plate and check the clutch centre height as shown

31.3a Remove wire circlip from centre of clutch drum ...

31.3b ... to release thrust bearing and shim washers

31.4a Assemble the plain washer, wave washer and thrust plate

31.4b Fit and tighten the clutch centre nut

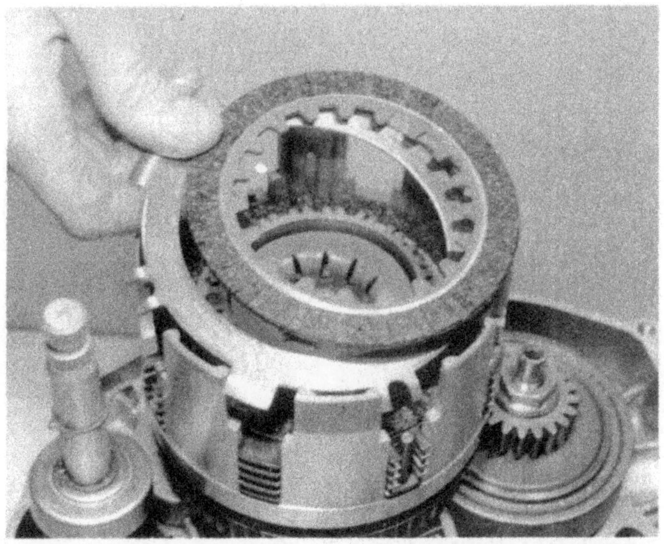

31.5 Fit the clutch plain and friction plates. Note position of plate with angled tangs

31.6 Check the clutch plate height as shown

31.7a Place the clutch springs over their support pins

31.7b Fit the pressure plate assembly, making sure that the circlip seats fully as shown

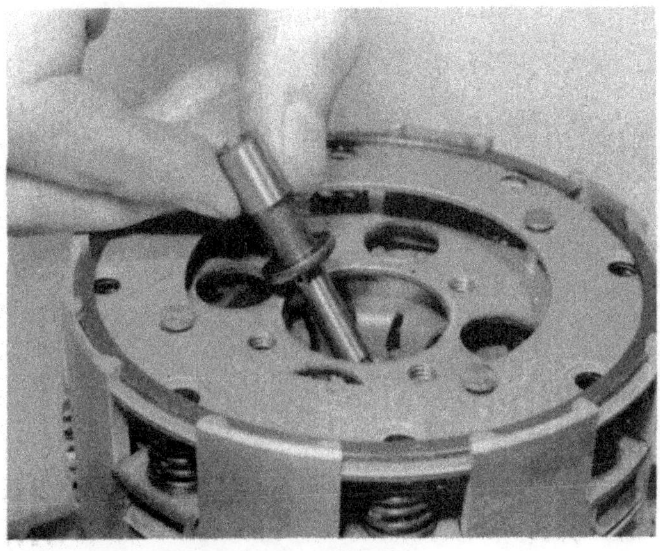

31.7c Check that the pushrods are in position ...

31.7d ... then offer up the bearing block

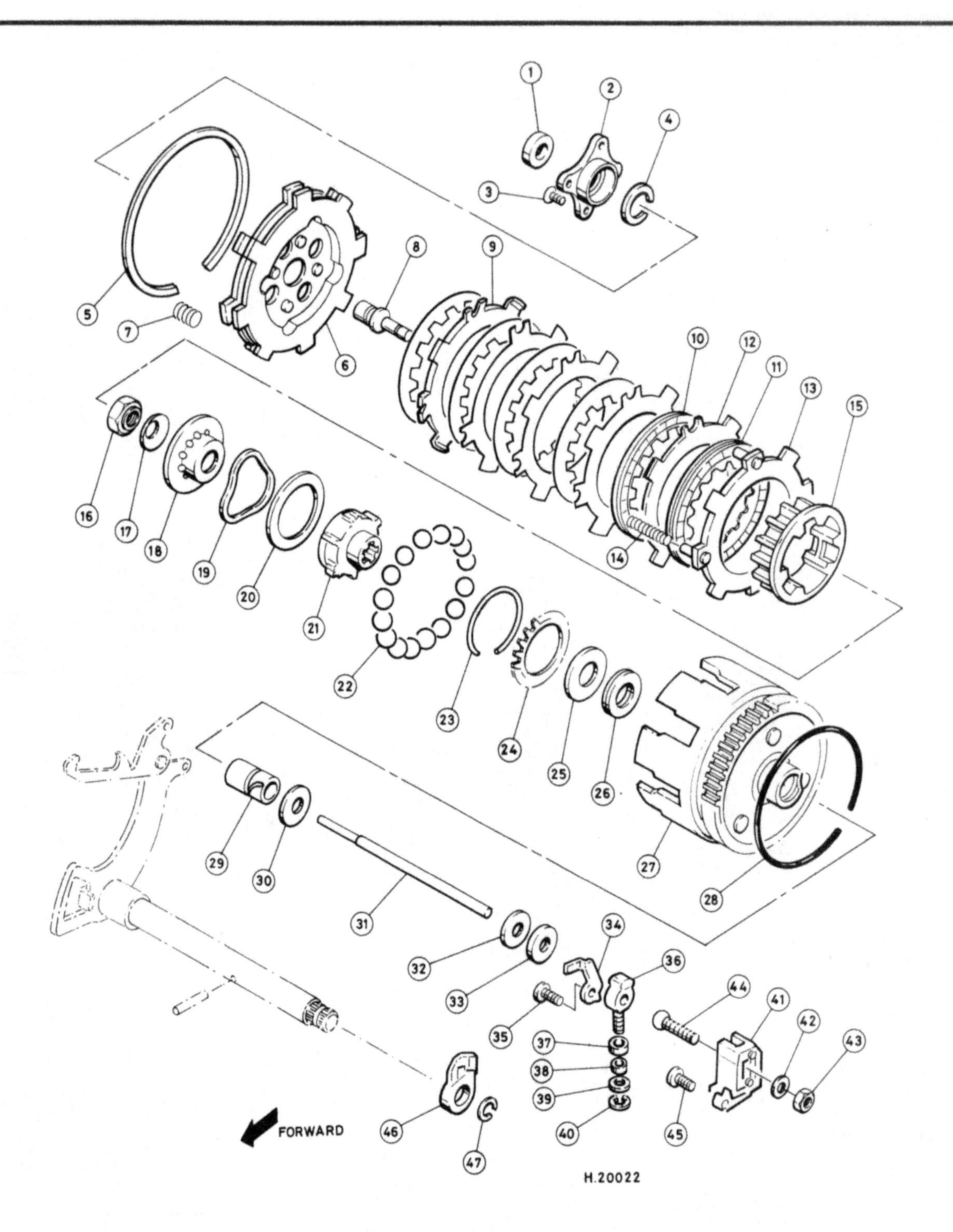

Fig. 1.13 Clutch

1 Bearing
2 Bearing block
3 Screw – 4 off
4 Circlip
5 Circlip
6 Pressure plate
7 Spring – 8 off
8 Short pushrod
9 Plain plate with angled tangs
10 Friction plate – 4 off/T50, 5 off/T80
11 Friction plate
12 Plain plate – 3 off/T50, 4 off/T80
13 Thrust weight plate
14 Spring – 4 off
15 Clutch centre
16 Nut
17 Washer
18 Thrust plate
19 Wave washer
20 Plain washer
21 One-way boss
22 Steel ball – 12 off/T50, 16 off/T80
23 Circlip
24 Toothed thrust washer – early model only
25 Shim
26 Thrust bearing
27 Outer drum
28 O-ring
29 Inner sleeve
30 Thrust washer
31 Long pushrod
32 Bush
33 Oil seal
34 Spring plate
35 Screw
36 Release rocker
37 Bush
38 Bush
39 Thrust washer
40 Circlip
41 Adjuster housing
42 Washer
43 Locknut
44 Adjusting screw
45 Screw
46 Cam plate
47 Circlip

31.8a Check the condition of the seals in the outer cover ...

31.8b ... then fit the cover, followed by the kickstart lever

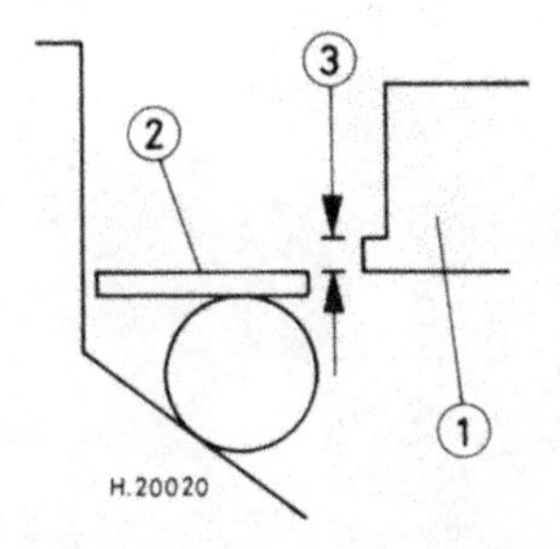

Fig. 1.14 Clutch centre height check

1 *Clutch centre*
2 *Thrust weight plate*
3 *Correct height – 0.2 – 0.4 mm (0.0078 – 0.0157 in)*

32 Reassembling the engine/transmission unit: refitting the piston and cylinder barrel

1 Position the crankshaft at TDC, so that the connecting rod is projecting as far as possible from the crankcase. Pack some clean rag around the connecting rod to support it and to exclude dirt or dropped circlips from the crankcase. Lubricate the connecting rod small-end eye, then offer up the piston, noting that the arrow mark must face downwards, towards the exhaust port.

2 Push the gudgeon pin into place, then fit **new** circlips to retain it. Note that the risk of a used circlip working loose is significant; never re-use old circlips. Make sure that the circlips seat fully and that the open end is positioned opposite the removal slots in the piston.

3 Check that the gasket face is clean, then fit the two dowel pins to the left-hand studs (nearest the cam chain tunnel). Place a new cylinder base gasket in position.

4 Position the ring ends at 120° intervals around the piston, and lubricate both the piston and the bore with clean engine oil. Slide the barrel over the holding studs until it reaches the piston. If a ring compressor is to be used, fit this over the piston, making sure that the rings are fully compressed, then tap the barrel down over the top half of the piston using the palm of the hand. Be careful that the ends of the oil ring expander do not overlap.

5 In the absence of a ring compressor, gradually work the barrel down the studs, compressing each ring in turn by hand, and guiding it into the tapered lead-in at the base of the bore. Make sure that each ring enters the bore fully, then slide the barrel down to the next ring and repeat the procedure. On no account use force; the rings are brittle and are easily snapped if care is not taken. Once all of the rings are engaged in the bore, remove the rag from the crankcase mouth and feed the cam chain through its tunnel in the barrel before pushing the barrel down onto its gasket.

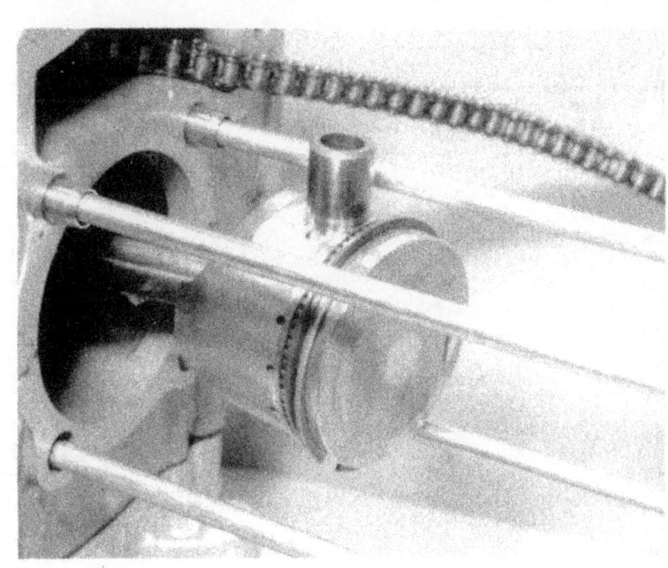

32.1a Refit the piston on the connecting rod ...

32.1b ... noting that arrow mark must face towards exhaust port

32.4 Fit the barrel over the piston, guiding the rings into the lead-in at the bottom of the bore

33.1a Fit the cam chain tensioner blades, positioning them as shown

33 Reassembling the engine/transmission unit: refitting the cylinder head

1 Check that the mating surfaces of the cylinder head and barrel are clean and free from oil. Fit the dowel pins over the left-hand studs, pushing them down to engage in their recesses on the top surface of the cylinder barrel. Place a new cylinder head gasket over the studs. Fit the cam chain tensioner blade and guide blade into their respective recesses, making sure that they seat correctly and are round the right way.

2 Offer up the cylinder head, passing the cam chain loop up through the tunnel in the head casting. Push the head down onto the gasket surface.

3 Fit the domed nuts to the cylinder head studs. Tighten the four nuts evenly and progressively, in a diagonal sequence, to the final torque figure of 1.0 kgf m (7.2 lbf ft). Fit and tighten the two bolts which secure the head to the barrel, on the left-hand side.

33.1b Fit a new gasket, and check that the dowels are in place

33.3 Tighten the cylinder head nuts evenly and progressively to the specified torque figure

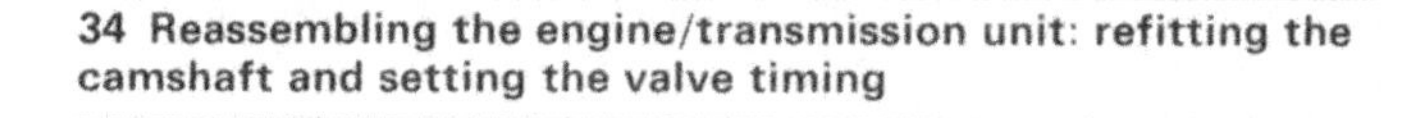

34 Reassembling the engine/transmission unit: refitting the camshaft and setting the valve timing

1 If the cam chain tensioner assembly is in place, remove the tensioner locking bolt from the side of the cylinder head to release pressure from the cam chain. Turn the alternator rotor until the 'T' mark aligns with the reference mark cast into the crankcase. When turning the rotor, hold the cam chain taut with a screwdriver or similar to prevent it from bunching up and jamming. If the camshaft is not already in place in the head, apply molybdenum disulphide grease to the cam lobes and slide it into position. Position the camshaft so that the lobes face towards the combustion chamber; ie, with the lobes away from the rocker ends. Align the slot in the end of the camshaft opposite the cast-in mark on the cylinder head. Slacken off the valve adjusters.

2 Fit the cam sprocket into the cam chain loop, positioning the alignment mark against the cast index mark in the cylinder head. Check that the rotor 'T' mark is still aligned, then offer up the sprocket, engaging the tang in the camshaft slot. Check that there is no chain slack on the exhaust side of the loop and that the cam sprocket and rotor 'T' marks align correctly. If the timing is out by a tooth, remove and refit the sprocket until the marks align correctly. Finally, refit the sprocket bolt and plain washer, and tighten the bolt to 2.0 kgf m (14 lbf ft).

3 Refit the tensioner plunger, spring and plug to take up the slack in the chain, tightening the plug to 1.5 kgf m (11 lbf ft). Using the alternator rotor, turn the crankshaft anticlockwise until the tang on the cam sprocket is in line with the tensioner locknut on the side of the cylinder head. Tighten the lockscrew firmly, securing the setting with the locknut. Refit and tighten the domed nut on the end of the lockscrew.

4 Turn the crankshaft anticlockwise through at least two complete revolutions and realign the 'T' mark. Check that the '0' mark on the cam sprocket aligns with its reference mark in the head casting. (If it is out by 180°, turn the crankshaft through one complete turn to align it.) If the mark does not align, check that the sprocket was positioned correctly (it may have been fitted one or more teeth out in relation to the chain). If the mark is out by less than one tooth, check that the camshaft chain drive sprocket is positioned correctly in relation to the keyway in the crankshaft, noting that this will entail the removal of the alternator rotor and stator to gain access. See Section 14 of this Chapter for further information.

5 Fit a new O-ring to the circular cylinder head cover, and place it in position on the cylinder head. Fit the two retaining bolts, tightening them to 0.7 kgf m (5.1 lbf ft). Remember that the valve clearances must be checked and adjusted before the engine is started.

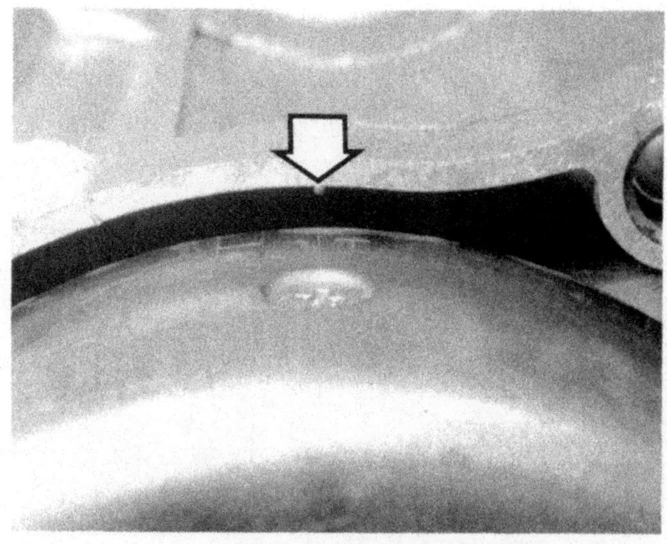

34.1a Align rotor 'T' mark with index mark on crankcase

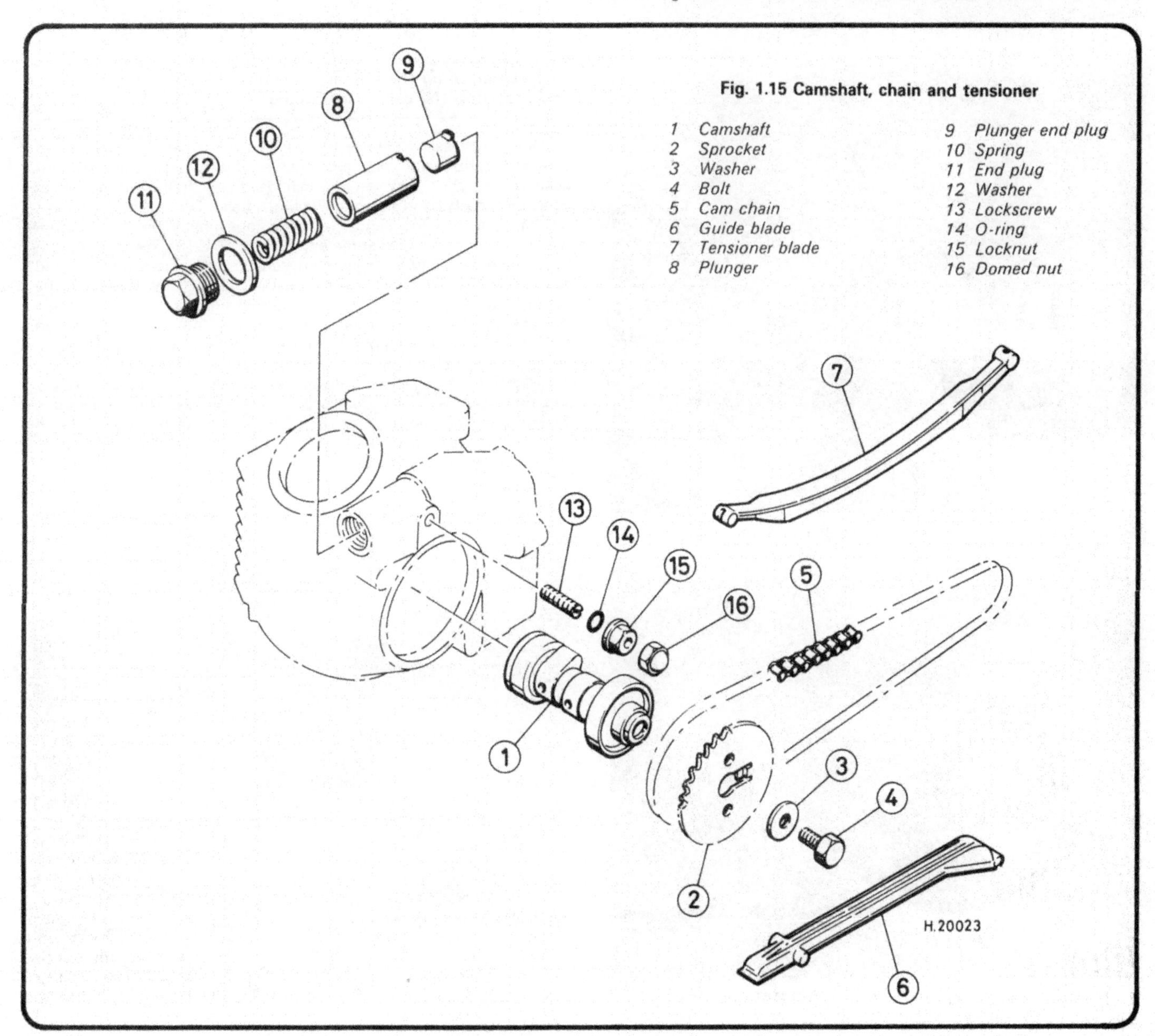

Fig. 1.15 Camshaft, chain and tensioner

1	*Camshaft*	9	*Plunger end plug*
2	*Sprocket*	10	*Spring*
3	*Washer*	11	*End plug*
4	*Bolt*	12	*Washer*
5	*Cam chain*	13	*Lockscrew*
6	*Guide blade*	14	*O-ring*
7	*Tensioner blade*	15	*Locknut*
8	*Plunger*	16	*Domed nut*

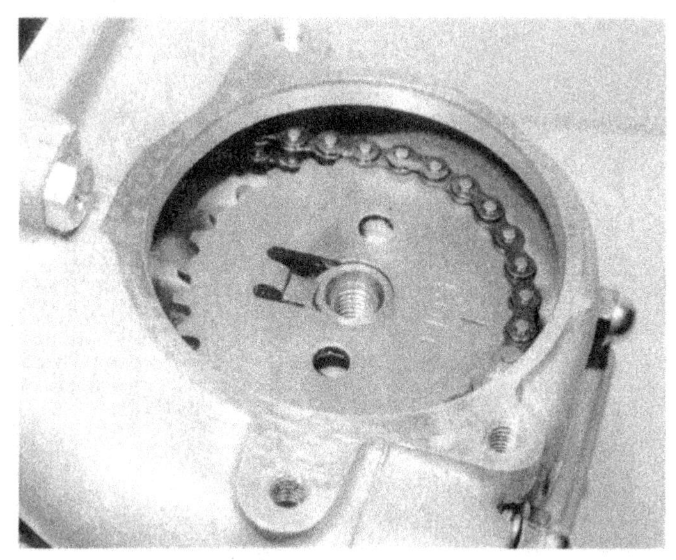

34.1b Fit sprocket, aligning timing mark with index mark at top of cylinder head

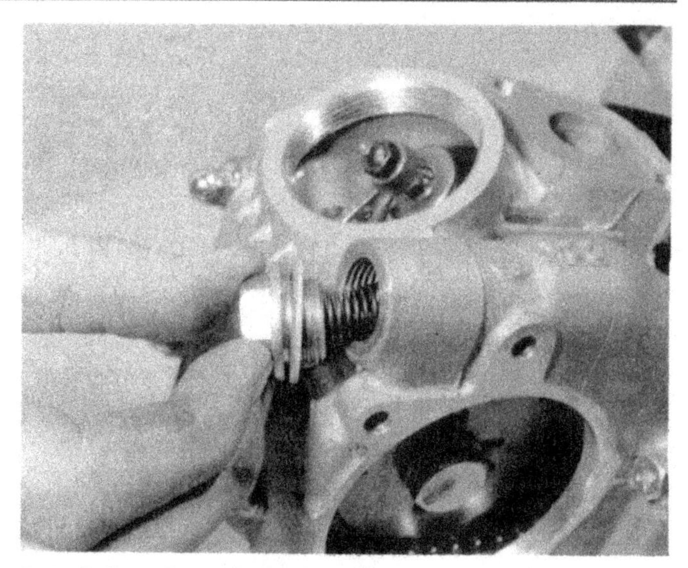

34.3a Refit tensioner plunger assembly ...

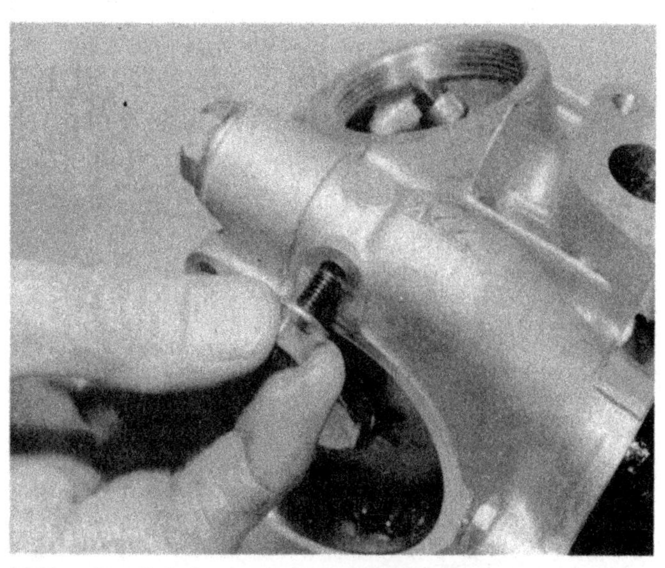

34.3b ... then fit lockscrew and set cam chain tension

35 Reassembling the engine/transmission unit: checking and resetting the valve clearances

1 Where necessary, remove the crankcase left-hand outer cover to expose the alternator rotor and unscrew the two circular hexagon-headed inspection caps on the cylinder head. Remove the circular cover on the left-hand side of the cylinder head. Turn the rotor until the 'T' mark on its edge aligns with the index mark on the adjacent crankcase edge. Check that the engine is at TDC (top dead centre) on the compression stroke. (The cam sprocket alignment mark should coincide with the reference mark on the head casting.) This can be established by grasping each rocker arm in turn by its hexagon head and pushing and pulling it. A small clearance should be detected on each one; if one seems tight or the valve partially open, turn the rotor through one full turn, realign the 'T' mark and check again.
2 Measure the clearance between each rocker arm adjuster screw and the top of its valve stem using feeler gauges. The clearance is indicated by the gauge which is a light sliding fit between the two components. Note the clearance and then repeat on the remaining valve. The specified figures are as follows:

Inlet .... 0.050 – 0.10 mm (0.002 – 0.004 in)
Exhaust .... 0.075 – 0.125 mm (0.003 – 0.005 in)

35.2 Use feeler gauges to set valve clearances

35.3 Refit inspection cover and valve covers

3 Where adjustment is required, and in the absence of the official tools, a suitable ring spanner can be used on the locknut, whilst the adjuster can be turned with the appropriate size of magneto spanner. Slacken the locknut by about 1/2 turn, holding it in this position whilst the adjuster is turned to give the required setting. Check the clearance with feeler gauges, then holding the adjuster, secure the locknut. Re-check the clearance and make any corrections required before repeating the procedure on the remaining valve. Finally, refit the inspection covers and the crankcase left-hand cover.

36 Reassembling the engine/transmission unit: refitting the carburettor

The carburettor and its inlet adaptor can be fitted at this stage if required, but may also be fitted after the engine/transmission unit has been installed. The fitting procedure is a simple reversal of the removal sequence and is dependent on whether the instrument was removed with or without the inlet adaptor. Check that all fasteners are tightened securely, and that the various pipe connections are remade.

37 Reassembling the engine/transmission unit: refitting the rebuilt unit into the frame

1 Position the engine unit below the frame. Have ready the two mounting bolts and their nuts. Both bolts are fitted from the left-hand side of the machine. As the unit is lifted into position it will be necessary to engage the universal joint from the middle gear assembly over the splines of the driveshaft. This is eased considerably if the rear wheel is turned to bring the splines into engagement.
2 In an emergency, the unit can be fitted single-handed, but it is much easier if one person lifts it into position while a second inserts the bolts; this approach is less likely to result in grazed knuckles or paintwork. Once the bolts are fitted loosely, the unit will hang in position in the frame, and the nuts can be fitted at leisure. Fit the engine mounting nuts, tightening them securely.
3 Refit the carburettor, or the carburettor top, depending on the method used during removal. Reconnect the fuel and vacuum pipes, and check that the throttle cable is correctly adjusted (see Routine Maintenance). Reconnect the spark plug HT lead.
4 Trace and reconnect the alternator and ignition pickup wiring at the connectors on the CDI unit and regulator/rectifier unit. The wiring is colour-coded to indicate the correct connections.
5 Fit a new gasket into the exhaust port, using a dab of grease to hold it in position. Offer up the exhaust system, fitting the exhaust port retainer and silencer mounting nuts loosely. Tighten the system down, starting with the exhaust port, then the silencer bracket nut.
6 Refit the kickstart lever, and also the footrest assembly if they were removed. Check that the crankcase drain bolt is secure, then fill the crankcase with the prescribed amount of oil; approximately 1.0 litre (1.8 Imp pint) of SAE 10W/40 type SE or SF motor oil. Note that the oil level must be checked after the engine has been run for the first time.
7 It is recommended that the engine should be started and run for a few minutes at this stage; it is easier to check around the unit for leaks or loose fittings before the legshield assembly is refitted. When refitting the legshield, tighten all fasteners finger-tight before final tightening. Check that the legshield is aligned correctly, then tighten the screws evenly. Beware of overtightening.

37.2a Lift engine into position, guiding driveshaft coupling over shaft splines

37.2b Fit and tighten the engine unit upper ...

37.2c ... and lower mounting bolts

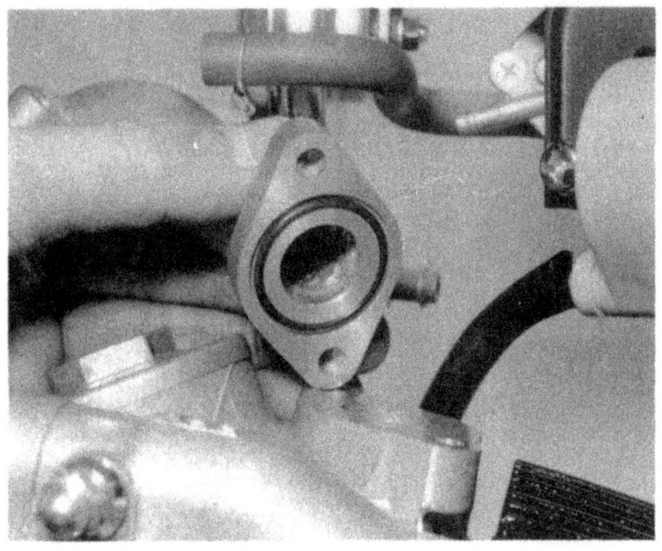

37.3a Refit the inlet adaptor and carburettor, using new O-rings

37.3b Reconnect the throttle and choke cables

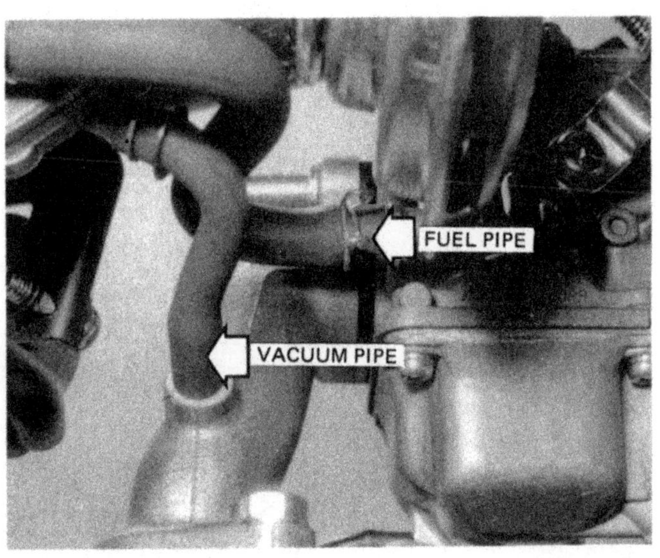

37.3c Refit the fuel and vacuum pipes

37.3d Connect inlet hose from air filter to carburettor

37.4a Reconnect alternator and ignition wiring at connectors behind the left-hand side panel

37.4b Install and reconnect the battery

37.5a Use a new sealing ring in the exhaust port – hold in place with grease

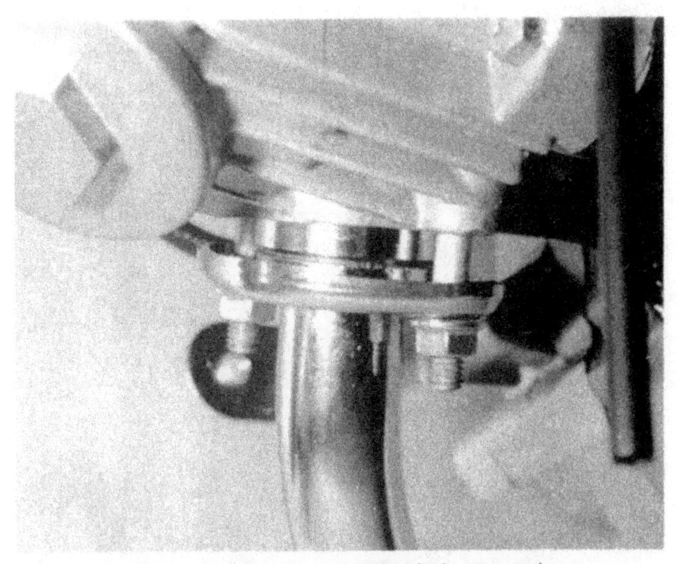

37.5b Refit the exhaust system, tightening it down evenly

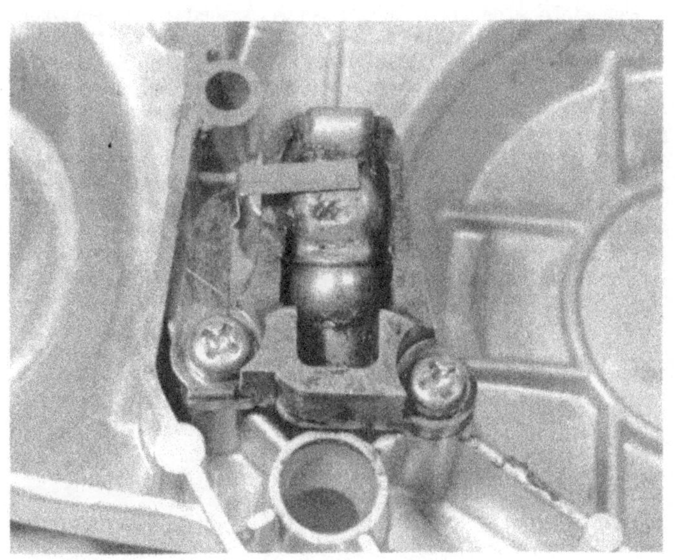

37.6a Grease the clutch release rocker assembly ...

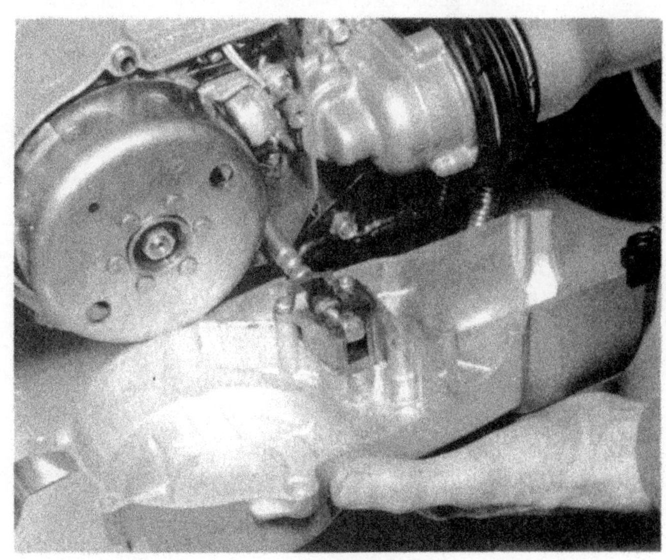

37.6b ... and refit the left-hand outer cover

37.6c Check and adjust the clutch pushrod (Routine Maintenance)

37.6d Refit the footrest assembly to underside of crankcase

37.6e Refill unit with specified quantity of SAE 10W/40 motor oil

38 Starting and running the rebuilt engine

1 Attempt to start the engine using the usual procedure adopted for a cold engine. Do not be disillusioned if there is no sign of life initially. A certain amount of perseverance may prove necessary to coax the engine into activity even if new parts have not been fitted. Should the engine persist in not starting, check that the spark plug has not become fouled by the oil used during reassembly. Failing this go through the fault finding charts and work out what the problem is methodically.
2 When the engine does start, keep it running as slowly as possible to allow the oil to circulate. Open the choke as soon as the engine will run without it. During the initial running, a certain amount of smoke may be in evidence due to the oil used in the reassembly sequence being burnt away. The resulting smoke should gradually subside.
3 Check the engine for blowing gaskets and oil leaks. Before using the machine on the road, check that all the gears select properly, and that the controls function correctly.

39 Taking the rebuilt machine on the road

1 Any rebuilt machine will need time to settle down, even if parts have been replaced in their original order. For this reason it is highly advisable to treat the machine gently for the first few miles to ensure oil has circulated throughout the lubrication system and that any new parts fitted have begun to bed down.
2 Even greater care is necessary if the engine has been rebored or if a new crankshaft has been fitted. In the case of a rebore, the engine will have to be run-in again, as if the machine were new. This means greater use of the gearbox and a restraining hand on the throttle until at least 500 miles have been covered. There is no point in keeping to any set speed limit; the main requirement is to keep a light loading on the engine and to gradually work up performance until the 500 mile mark is reached. These recommendations can be lessened to an extent when only a new crankshaft is fitted. Experience is the best guide since it is easy to tell when an engine is running freely.
3 If at any time a lubrication failure is suspected, stop the engine immediately, and investigate the cause. If any engine is run without oil, even for a short period, irreparable engine damage is inevitable.
4 When the engine has cooled down completely after the initial run, recheck the various settings, especially the valve clearances. During the run most of the engine components will have settled into their normal working locations. Check the various oil levels, particularly that of the engine as it may have dropped slightly now that the various passages and recesses have filled. Refer to Routine Maintenance and check the clutch adjustment.

Chapter 2 Fuel system and lubrication

Contents

Specifications

Fuel tank

Capacity	5.0 litre (1.1 Imp gal)

Fuel grade

Fuel grade	95 RON octane unleaded, or leaded

Carburettor

	T50	T80
Make	Mikuni	Mikuni
Type	VM14SH	VM16SH
Identification mark	35T-00	46J00
Main jet	77.5	85
Air jet	1.2	1.2
Jet needle (clip position)	3N8-4	3N8-3
Needle jet	E-1	E-4
Pilot jet	12.5	15
Pilot air screw setting	1¾ turns out	1⅝ turns out
Pilot outlet	0.7	0.7
Float valve seat	1.2	1.2
Fuel level	3 – 4 mm (0.118 – 0.157 in)	3.5 ± 1 mm (0.138 ± 0.004 in)
Idle speed	1700 ± 100 rpm	1700 ± 100 rpm
Inlet vacuum at idle speed	270 mm (10.63 in) Hg	270 mm (10.63 in) Hg

Engine/transmission lubrication

System type	Forced, wet sump
Pump type	Trochiod
Pump clearances – T50:	
Inner rotor tip to outer rotor tip	0.15 mm (0.006 in)
Service limit	0.2 mm (0.008 in)
Rotor endfloat	0.06 – 0.10 mm (0.0024 – 0.0039 in)
Service limit	0.15 mm (0.006 in)
Pump clearances – T80:	
Inner rotor tip to outer rotor tip	0.15 mm (0.006 in)
Rotor endfloat	0.06 – 0.10 mm (0.0024 – 0.0039 in)
Outer rotor to body	0.13 – 0.18 mm (0.0051 – 0.0071 in)
Filtration type	Pickup strainer
Oil capacity:	
At oil change	0.85 litre (1.5 Imp pint)
Dry	1.0 litre (1.8 Imp pint)
Oil grade	SAE 10W/40, type SE or SF motor oil

Torque wrench settings

Component	kgf m	lbf ft
Carburettor inlet adaptor	0.7	5.1
Carburettor	0.7	5.1
Air filter cover	0.7	5.1
Air filter casing	0.7	5.1
Oil pump cover	0.4	2.9
Oil pump mounting	0.7	5.1
Oil drain plug	2.0	14.0

1 General description

The fuel system comprises the fuel tank, located below the seat, from which fuel is gravity-fed to the fuel tap and carburettor. The tap is mounted next to the carburettor and has 'ON' and 'PRI' settings only; The tap is controlled by engine vacuum while set to the 'ON' position, the 'PRI' or prime setting being provided to allow the carburettor float bowl to be filled after running dry. From the tap, fuel flows through an inlet filter into the carburettor float bowl, where it is maintained at the correct level by the float-operated valve.

The carburettor is of the conventional concentric design, the fuel and air flow, and thus the engine speed, being controlled by the throttle valve, the needle jet and needle, and by the various jet sizes. Cold starting is aided by a butterfly-type choke, controlled by a handlebar-mounted lever.

Air is drawn into the carburettor via an oil-impregnated foam air filter element housed in a plastic filter box just below the steering head. The filter removes any airborne dust before it can enter the engine and accelerate wear.

Engine lubrication is catered for by a pump-fed lubrication system. The trochoid pump draws oil through a gauze strainer from the sump area of the crankcase. From here it is fed under pressure to the highly loaded engine components such as the crankshaft assembly, camshaft and rockers, the residual oil draining back to the bottom of the crankcase where the cycle is repeated.

In addition to the pickup strainer, a centrifugal oil filter is fitted on the right-hand end of the crankshaft. It is designed to remove the smaller particles which pass through the strainer. Oil is fed from the crankcase right-hand outer cover into the end of the crankshaft. Before the oil reaches the crankshaft it must pass through the spinning sludge trap, where centrifugal forces throw the contaminants to its outer edges, leaving clean oil to be fed to the bearings. The sludge trap needs to be cleaned out periodically to remove accumulated debris.

2 Fuel tank: removal, flushing and refitting

1 The fuel tank will not normally need to be removed unless the fuel has become contaminated with dirt or water, in which case it should be taken off and thoroughly flushed out. It is preferable to remove the legshield assembly before work starts (see Chapter 4). Because of the unavoidable spillage of fuel, this next stage is best done outside, well away from any possible sources of fire.

2 Have ready a container of at least 1 Imp gallon capacity. Disconnect the fuel pipe at the carburettor end by squeezing together the 'ears' of the retaining clip and sliding it along the pipe. The pipe can now be worked off its stub using a small screwdriver. Note that as soon as the pipe comes free of the stub, fuel will begin to run out of the pipe from the tank. Place the pipe end in the can and allow the fuel to drain fully.

2.3b ... to gain better access to front mounting bolts

3 Unlock and open the seat. Remove the two bolts which secure the seat hinge and the front of the fuel tank, and lift the seat away. Remove the two bolts at the rear of the tank. Trace and disconnect the leads to the fuel gauge sender unit (the connectors are hidden inside the frame pressing and can be located by pulling the leads up through the grommet through which they pass). Before the tank is lifted out, note the routing of the fuel pipe. This can be either removed along with the tank, or the tank can be lifted slightly and the pipe disconnected at the tank outlet. The latter method leaves the pipe routed out through the frame and may prove marginally easier during installation.

4 Once the tank has been removed, check it carefully for signs of dirt or water contamination. Of the two, water causes far more problems, and even a small blob can cause annoying jet blockages. Flush the tank with clean petrol, disposing of the waste fuel safely. Only when the tank is known to be quite clean should it be refitted.

5 In the event of tank leakage, have it repaired professionally. It is possible to repair steel fuel tanks by welding, but for obvious reasons there is a risk of fire or explosion if this is not tackled correctly. **Do not** attempt to weld, braze or solder the tank at home.

6 When refitting the tank, check that the fuel pipe is routed correctly and is fitted securely to the tank and fuel tap stub. Remember that a slight leak from the tank stub end of the pipe is unlikely to be noticed, but it will waste fuel and pose a serious fire risk. Do not omit to reconnect the fuel gauge sender leads.

2.3a Seat latch assembly can be removed ...

2.3c Right-hand rear bolt also secures seat stay

2.3d Disconnect outlet pipe from tank stub ...

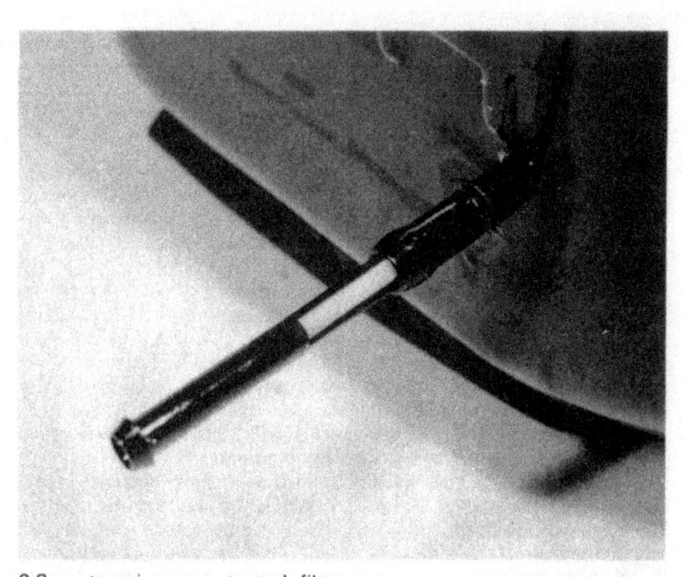

2.3e ... to gain access to tank filter

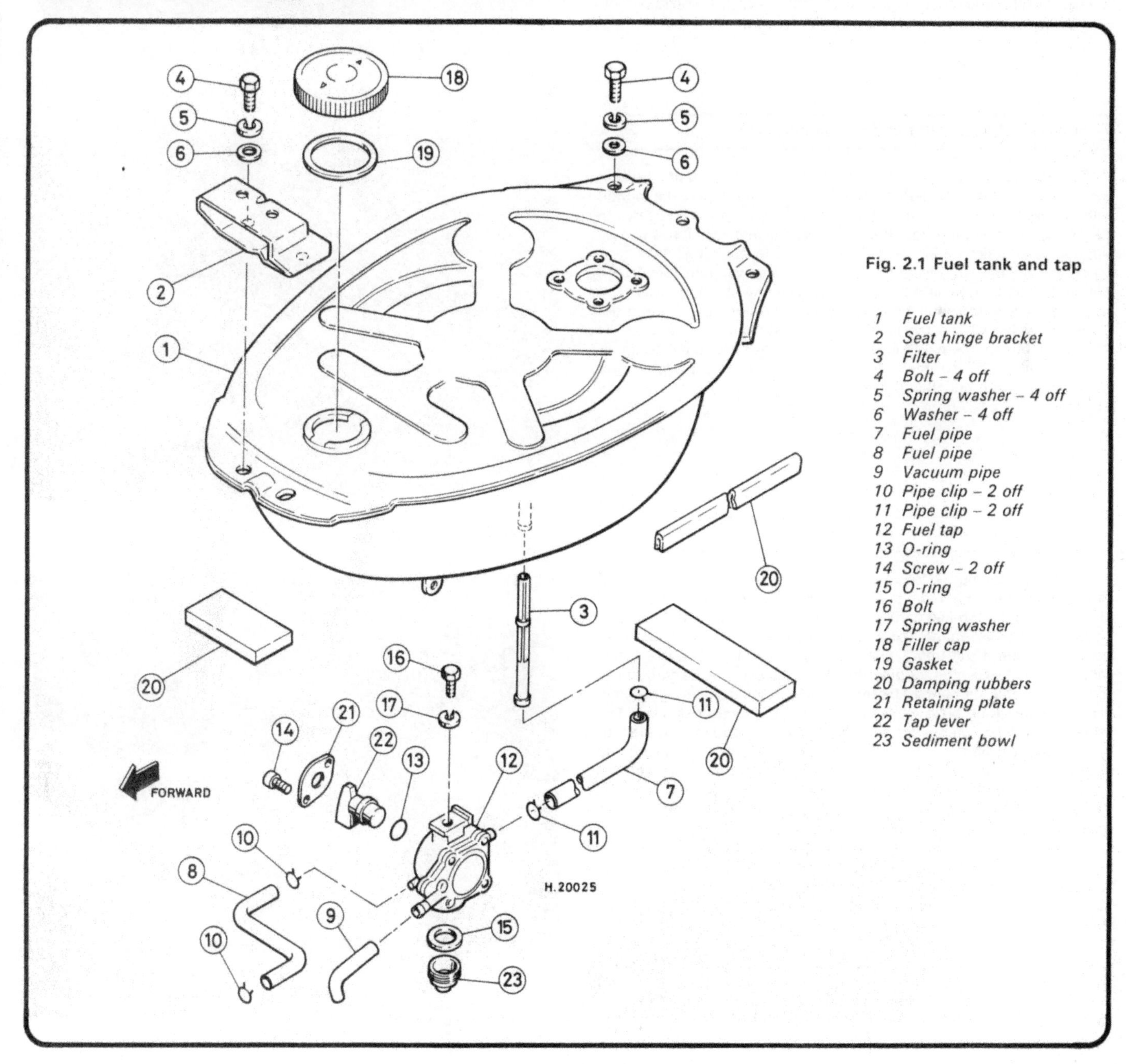

Fig. 2.1 Fuel tank and tap

1 Fuel tank
2 Seat hinge bracket
3 Filter
4 Bolt – 4 off
5 Spring washer – 4 off
6 Washer – 4 off
7 Fuel pipe
8 Fuel pipe
9 Vacuum pipe
10 Pipe clip – 2 off
11 Pipe clip – 2 off
12 Fuel tap
13 O-ring
14 Screw – 2 off
15 O-ring
16 Bolt
17 Spring washer
18 Filler cap
19 Gasket
20 Damping rubbers
21 Retaining plate
22 Tap lever
23 Sediment bowl

3 Fuel and vacuum pipes, examination and renewal

1 The fuel feed pipe should be checked regularly for signs of splitting, perishing or other damage. Bear in mind that the fuel tap is located on the frame near to the carburettor, not at the tank. This means that if the pipe leaks, the entire contents of the tank will be lost, creating a considerable fire risk.
2 The pipe is made from thin-walled synthetic rubber, and must be replaced with a pipe made from the same material. On no account use natural rubber tubing, which is soluble in petrol, producing a sludge which will quickly block the carburettor jets. Clear plastic tubing is best avoided too; the plasticiser in this material will be leached out eventually; leaving the pipe brittle and prone to splitting.
3 Check that the fuel and the various drain and breather pipes are secure and that their retaining clips are in position.
4 The vacuum pipe (the smaller of the two pipes connected to the tap) can be dealt with in the same way as has been described for the fuel pipe. Note that a split or hole in the pipe may result in the fuel tap not opening when the engine is started (if this happens on the road, try switching to the 'PRI' setting until repairs can be made). The resulting air leak into the inlet adaptor will also upset the fuel/air mixture.

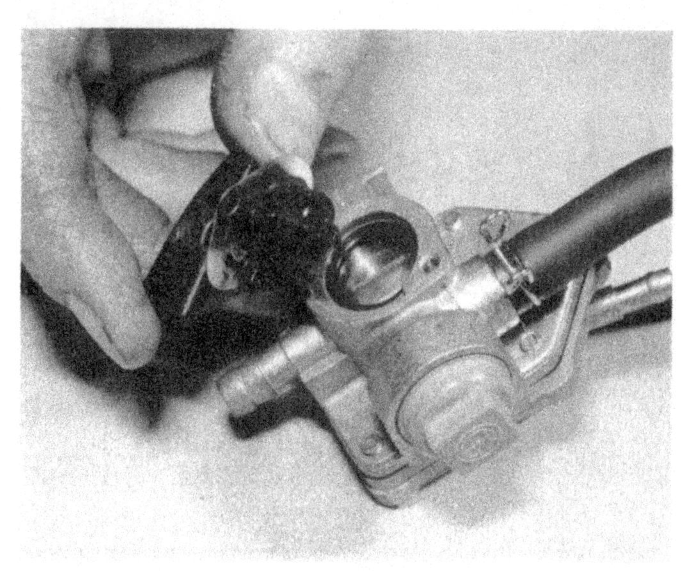
4.2a Tap lever assembly can be detached to renew O-ring

4 Fuel tap: dismantling and reassembly

1 The fuel tap, as has been mentioned, is attached to a frame bracket very close to the carburettor. It can be removed for examination without removing the carburettor. It will first be necessary to disconnect the fuel pipe from the tank, and if care is taken, this can be plugged to contain the fuel to avoid the need for draining.
2 To remove the tap, disconnect the remaining (vacuum) pipe, then remove the single retaining bolt. The tap can now be lifted away. The most likely cause of problems is a leak between the tap body and the O-rings which seal the tap lever and the sediment bowl. These are renewable, but an internal fault such as a damaged vacuum diaphragm poses something of a problem. No replacement parts are available, and the diaphragm can only be obtained as an assembly, together with the tap body.

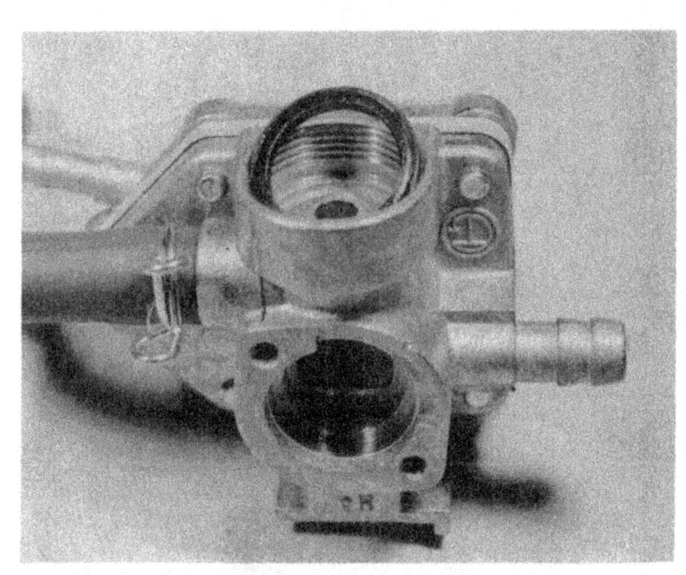
4.2b Sediment bowl is sealed by O-ring

5 Carburettor: removal and refitting

1 Start by removing the legshield assembly as described in Chapter 4. Check that the fuel tap is set to the 'ON' position, then prise off the fuel pipe between the tap and carburettor. The vacuum pipe should also be disconnected. Disengage the throttle cable at the carburettor end, lodging it around the frame, clear of the carburettor. Free the choke cable in a similar fashion, turning the actuating lever on the carburettor so that the cable nipple can be disengaged. Slacken the single screw which secures the choke cable outer to the carburettor bracket. The cable can then be disengaged from the choke arm and lodged clear of the carburettor. Free the air filter connecting hose by releasing the retaining clip. Remove the two carburettor mounting bolts and lift away the carburettor and its heat insulator.
2 When refitting the carburettor, use a new O-ring between the carburettor and inlet adaptor heat insulator, if it appears flattened or damaged. Check that the mounting bolts are fitted securely, but beware of overtightening since the carburettor and inlet adaptor flanges are easily distorted. Reconnect the choke cable, and check that the choke operates normally. Refit the carburettor to air filter hose, making sure that the retaining clip is tightened securely. Refit the throttle cable and check throttle operation. Connect the fuel and vacuum pipes, and also the drain hose, making sure that it is routed correctly. Turn the fuel tap to the 'PRI' position to fill the float bowl, then return it to the 'ON' position and start the engine: allow it to stand for a few minutes before checking for leaks. Check the throttle cable and carburettor adjustments. If all is well, refit the legshield assembly.

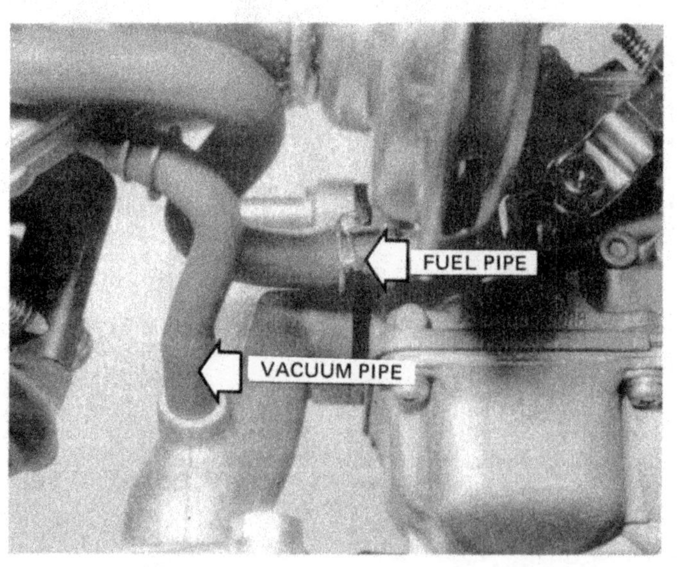

5.1a Disconnect the fuel and vacuum pipes

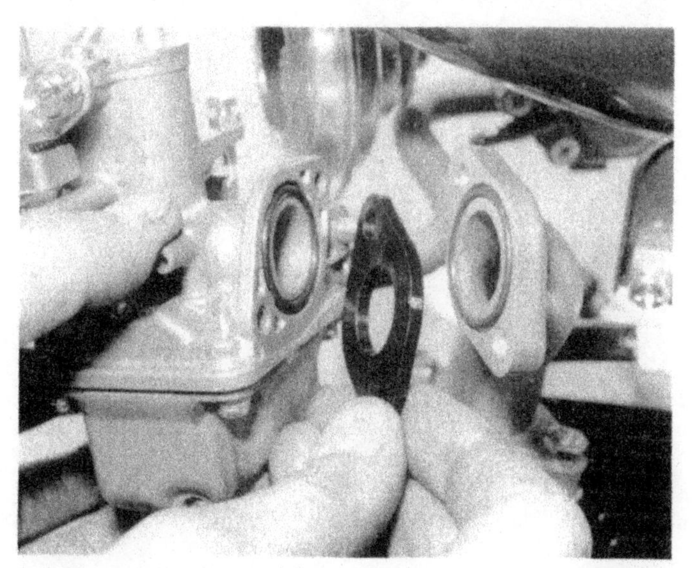

5.1b Note heat insulator and O-rings between carburettor and inlet adaptor

6 Carburettor: dismantling, examination and reassembly

1 Remove the four screws which retain the float bowl to the underside of the carburettor and lift it away. Displace the float pivot pin, and lift away the float, together with the valve needle.
2 From the centre of the carburettor body, unscrew the main jet from the longer of the projections of the carburettor body. The needle jet, which shares the same bore in the carburettor, can be checked and cleaned in position. The pilot jet, or slow running jet, is housed in the adjacent bore, and this can be unscrewed for cleaning.
3 Screw the pilot air and throttle stop screws inwards until they just seat, counting the number of turns and part turns until they stop (be careful not to overtighten them or the ends will be distorted). Make a note of these settings so that they can be refitted to the same positions during assembly. Both screws can now be removed for examination.
4 Moving to the upper half of the carburettor, remove the three screws which secure the combined carburettor top and throttle cable bracket and lift it away. Remove the accelerator pump lever (single cross-head screw), followed by its operating link which is attached to the throttle valve arm shaft by a single nut. Release the single screw which retains the throttle valve arm to its shaft, then displace and remove the shaft assembly.
5 The arm can now be lifted out of the carburettor, attached to the throttle valve assembly. Carefully unhook the arm from the throttle valve, taking care to avoid losing the spring which will tend to fly out. To release the jet needle, remove the two cross-head screws which secure the retainer to the top of the throttle valve. Carefully remove the accelerator plunger cap and displace the plunger from the carburettor body.
6 Examine the carburettor body and the float bowl for cracks or other damage. If damage is found, renew the affected part; it is not possible to repair successfully the die castings used for carburettors. Clean the carburettor castings carefully, blowing through the various drillings and passages with compressed air.
7 Check the jets for dirt or other obstructions. These should be cleared using compressed air; never attempt to push wire through the jet drillings or they will be ruined. In an emergency, it is permissible to use a fine nylon bristle or similar to dislodge any stubborn dirt, but avoid scoring or enlarging the fine drillings at all costs.
8 Examine the float for leakage by shaking it; any fuel inside the float will be quite obvious. If the float is leaking this will effect the effective float height, and the float should be renewed. Check the float needle valve for wear. If present, this will take the form of a visible ridge around the sealing face of the needle, and will tend to cause flooding. Renew the needle if it is suspect.
9 After a considerable mileage has been covered, the throttle valve, and possibly the carburettor body, will wear, allowing air to leak past the valve and upsetting the mixture. Scoring of the body or the valve is usually due to the engine being run with a damaged or missing air filter. If renewal of the throttle valve does not resolve the problem, a new body will be required.
10 The choke butterfly should not give rise to problems, and it should not normally be disturbed. Note that the choke butterfly valve and spindle are not available as replacement parts, implying that the carburettor body must be renewed complete if there are problems in this area.
11 Reassemble the carburettor by reversing the dismantling sequence, using the accompanying photographs and line drawing for guidance. Use new O-rings and gaskets throughout, and renew the throttle arm shaft seals. The shaft should be greased prior to installation. Make sure that all parts are clean, and be careful not to overtighten the jets or screws. In particular, do not overtighten the air screw, or it will be damaged. When fitting the accelerator pump plunger, take care to avoid scoring it, and do not omit its spring. Also ensure that the small gaiter fitted to the top part of the plunger is correctly located in its groove in the plunger. After the carburettor has been refitted, check the adjustments as described later in this Chapter.

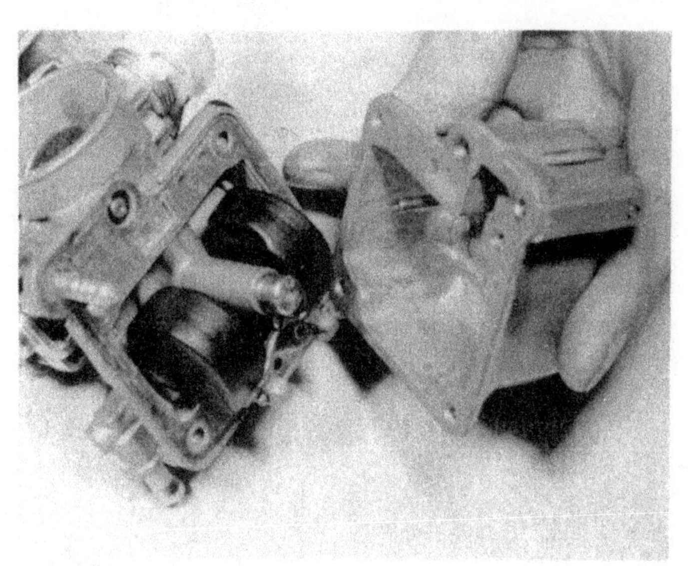

6.1a Remove float bowl screws and lift it away ...

6.1b ... taking care not to lose plunger spring

6.1c Displace the float pivot pin ...

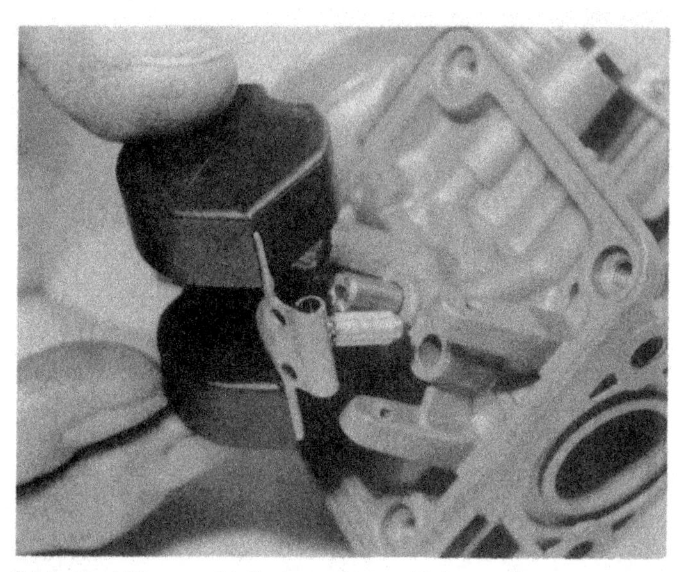

6.1d ... and lift away the float, complete with its needle

6.2a Pilot jet can be unscrewed for cleaning ...

6.2b ... as can the main jet

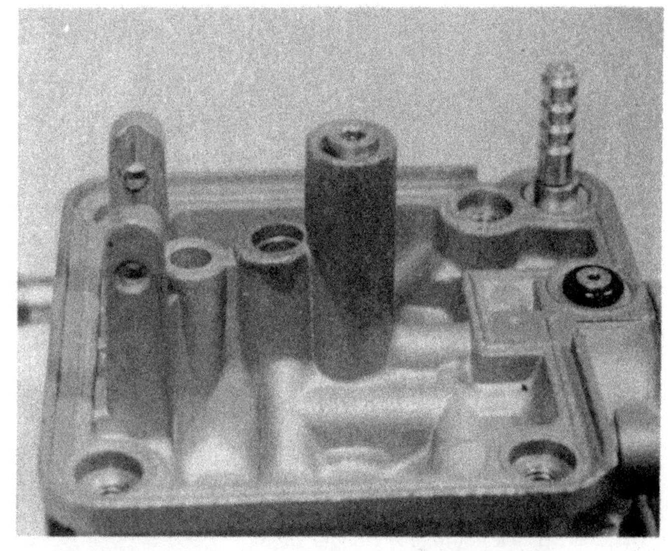

6.2c Needle jet remains in position in the central projection

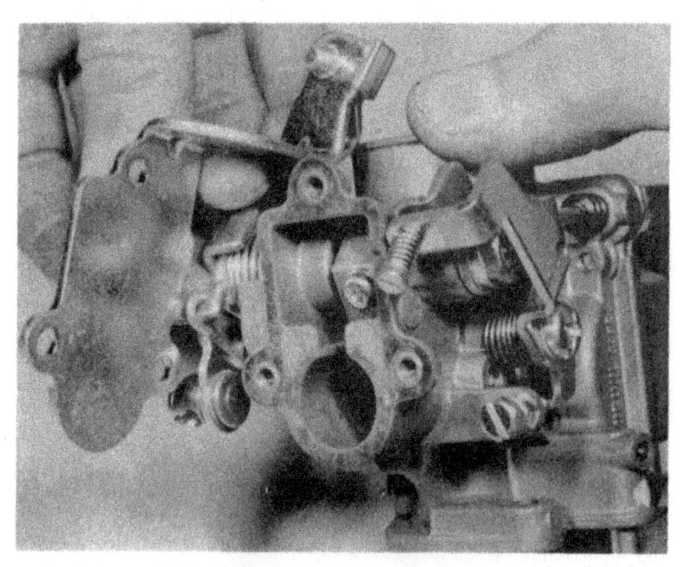

6.4a Carburettor top is secured by three screws

6.4b Remove accelerator pump lever (A) and operating link (B)

6.4c Throttle valve assembly is secured to its shaft by a single screw (arrowed)

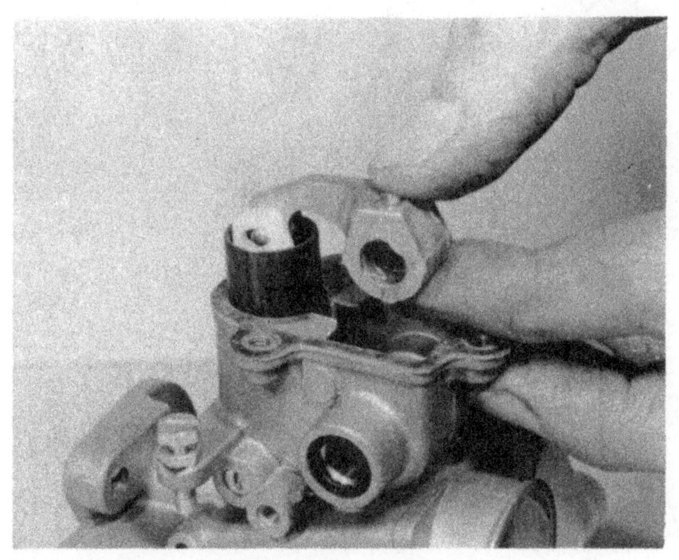
6.4d Withdraw the shaft and lift throttle valve clear

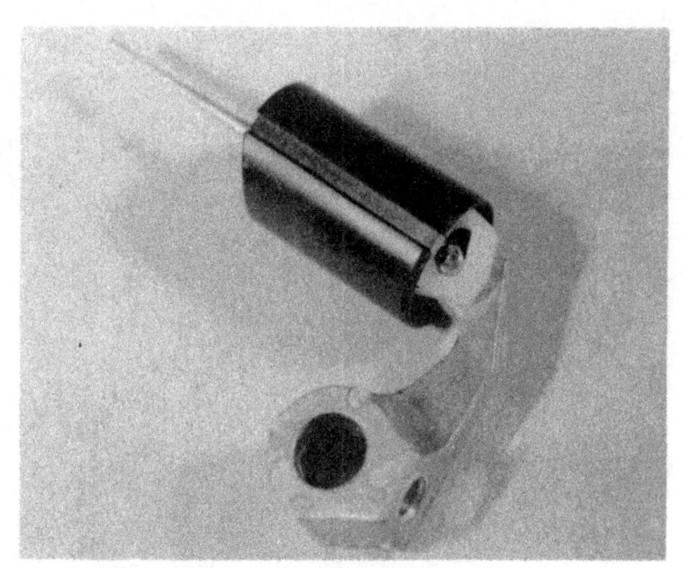
6.5a Unhook the arm from the throttle valve

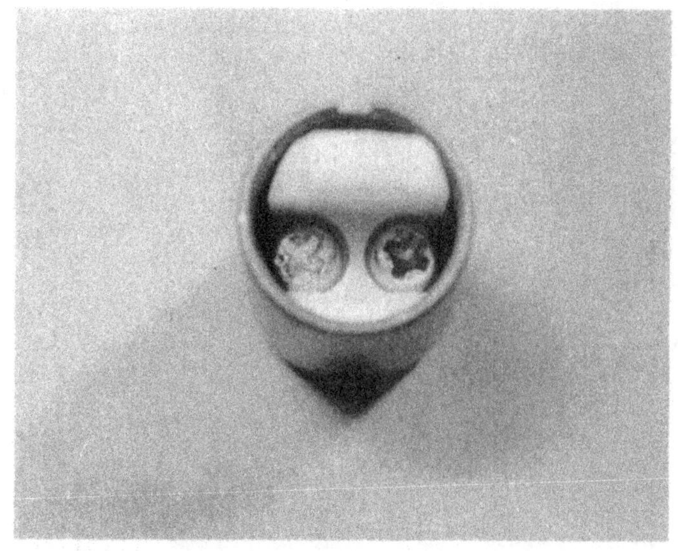
6.5b Remove the two screws which secure the retainer ...

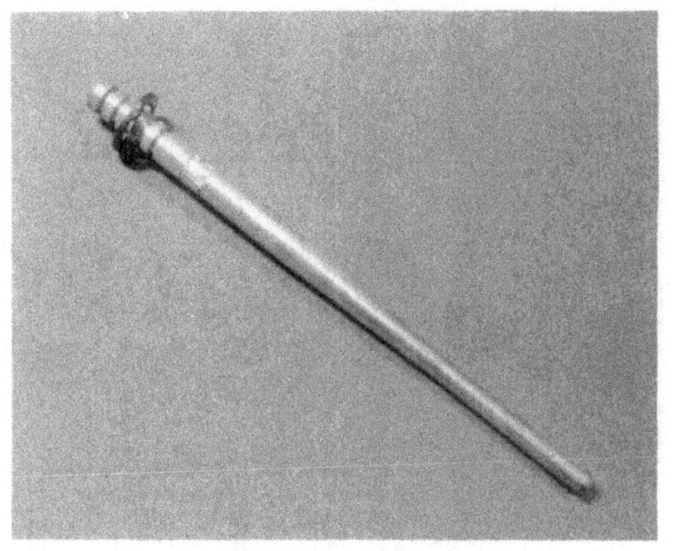
6.5c ... to free the jet needle from the throttle valve

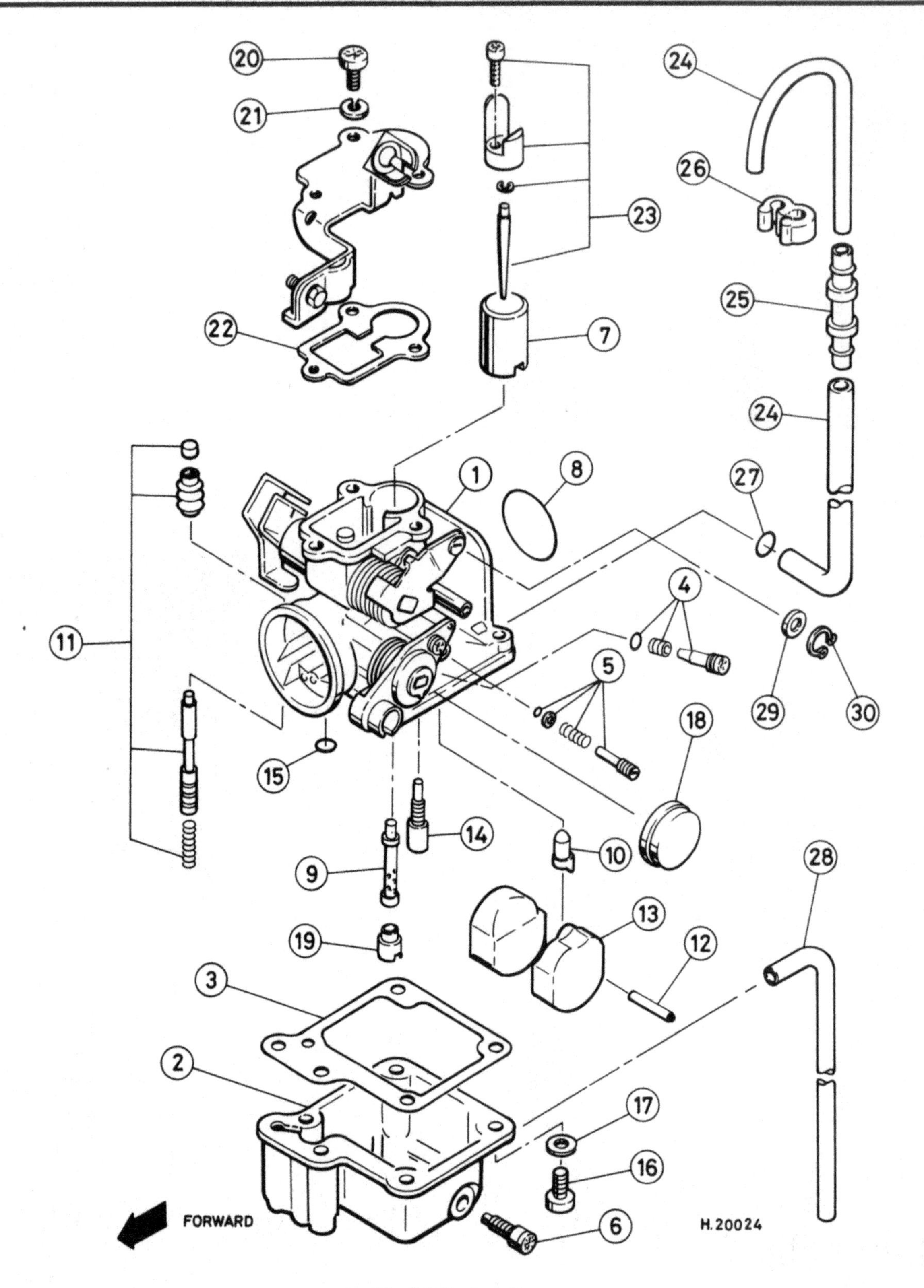

Fig. 2.2 Carburettor

1 Carburettor body
2 Float bowl
3 Gasket
4 Pilot air screw assembly
5 Throttle stop screw assembly
6 Drain screw
7 Throttle valve
8 O-ring
9 Needle jet
10 Float valve needle
11 Accelerator pump operating rod
12 Pivot pin
13 Float
14 Pilot jet
15 O-ring
16 Screw – 4 off
17 Spring washer – 4 off
18 Cap
19 Main jet
20 Screw – 3 off
21 Spring washer – 3 off
22 Gasket
23 Jet needle assembly
24 Fuel pipe
25 Pipe union
26 Pipe clamp
27 Pipe clip
28 Overflow pipe

6.11a Use new O-rings when rebuilding the carburettor ...

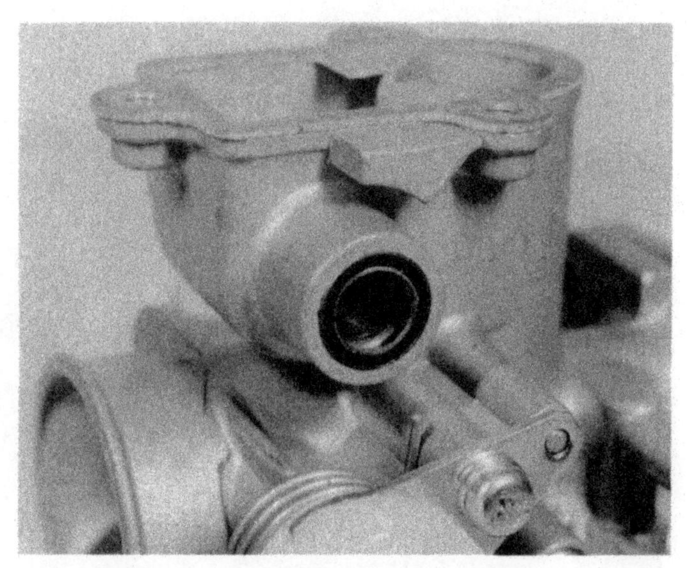

6.11b ... and renew the shaft seals to prevent air leaks

7 Carburettor: settings and adjustment

Jet sizes

1 These are determined during manufacture and are designed to suit the machine to which they are applied. They take into account operating conditions and the effects of the air filter and exhaust system, and thus should not need changing during the life of the machine. After an extremely high mileage has been covered it is possible that the original jets will have worn due to the constant erosive effect of the fuel passing through them, in which case fitting new jets of the specified sizes may restore performance and fuel economy. The most likely reason for incorrect mixture strength, however, is a choked air filter or exhaust system, or air leaks into the carburettor due to worn or damaged seals or O-rings. Always check these first before suspecting the carburettor jets.

Fuel level

2 This should be checked if there is reason to suspect the carburettor of producing an excessively weak or rich mixture. You will need about one foot or so of a small bore plastic tubing (the type used for car windscreen washers will suffice). Attach the tubing to the stub on the bottom of the float bowl, bringing the open end of the tube up alongside the float bowl. Slacken the float bowl drain screw, allowing the fuel to fill the tube, indicating the level in the float bowl. Check that the machine is level by comparing the indicated level in relation to the float bowl gasket face on each side of the machine. If necessary, pack the centre stand to level the machine.

3 Check that the fuel tap is set to 'ON', then start and run the engine for a few minutes to establish the normal running level. Now measure the height of the fuel in the tube below the lower edge of the carburettor body. This should be as shown in the Specifications. If incorrect, remove the float bowl and check the float needle tip for damage, renewing it if it appears worn. If the needle is in good condition, adjust the needle position in relation to the float by gently bending the tang to which it is attached. Note that only a very small change will be required. Reassemble the float bowl and recheck the level as described above.

Throttle cable adjustment

4 The throttle cable adjustment should be checked whenever the throttle valve or linkage, or the twistgrip unit have been disturbed. Check that there is 2 – 5 mm (0.08 – 0.20 in) free play, measured at the edge of the throttle twistgrip flange, before the throttle valve begins to move. If adjustment is required, slacken the locknut on the adjuster below the twistgrip, and set the adjuster to give the prescribed amount of free play, then secure the locknut.

Idle speed and pilot air screw adjustment

5 This setting must be maintained at all times. Note that it is of particular importance on any machine with an automatic clutch: too high an idle speed will lead to the machine 'creeping' forward any time a gear is engaged, and is potentially dangerous. The operation must be carried out with the engine at full operating temperature, preferably after a ride of several miles.

6 Start by setting the pilot air screw to its standard position of 1 5/8 turns out. (**Do not** screw the air screw hard against its seating – it will be damaged.) Start the engine with the machine on its stand and the rear wheel clear of the ground. To set the idle speed accurately, a test tachometer is needed. Failing this, try to obtain the lowest reliable idle speed.

7 Set the idle speed to 1700 ± 100 rpm (or the lowest reliable speed) using the throttle stop screw. Moving to the adjacent pilot air screw, try moving this in or out by 1/4 turn at a time until the most even idling is obtained. Reduce the idle speed once more, as required. Note that it will not be possible to obtain a reliable tickover if the centrifugal clutch is dragging; see Chapter 1 for details.

Accelerator pump adjustment

8 With the carburettor removed from the machine, check that the pump circuit operates when the throttle valve reaches the 1/3 open position. If not, adjust the screw (see photograph) until normal operation is restored. Note that if the circuit fails completely, dismantle the carburettor and clean all passages and drillings.

7.6 Set the pilot air screw to its standard setting (throttle stop screw A arrowed)

7.8 Use adjuster screw (arrowed) to regulate accelerator pump setting

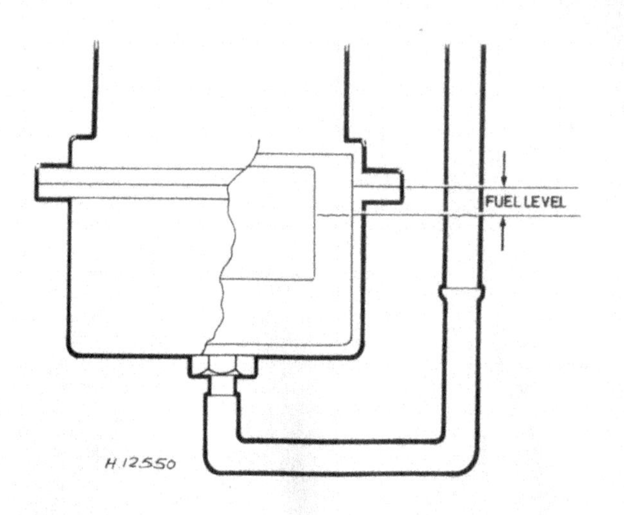

Fig. 2.3 Fuel level measurement

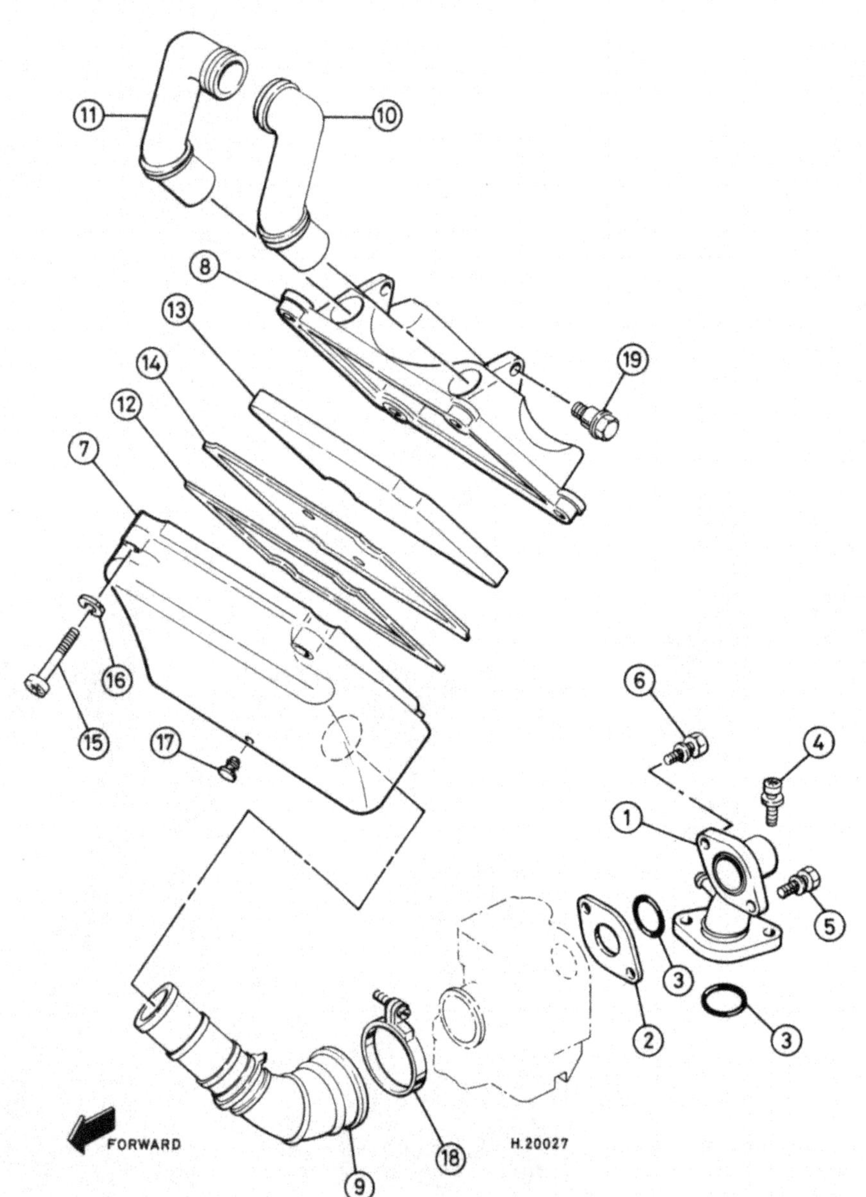

Fig. 2.4 Air filter and inlet adaptor

1 *Inlet adaptor*
2 *Heat insulator*
3 *O-ring – 2 off*
4 *Screw and washer – 2 off*
5 *Screw and washer*
6 *Screw*
7 *Air filter casing*
8 *Casing cover*
9 *Outlet hose*
10 *Left-hand air inlet hose*
11 *Right-hand air inlet hose*
12 *Seal*
13 *Element*
14 *Element support frame*
15 *Screw – 4 off*
16 *Washer – 4 off*
17 *Plug*
18 *Clamp*
19 *Bolt – 2 off*

8 Air filter element: removal and cleaning

1 This operation should be carried out without fail at the intervals specified in Routine Maintenance, and more frequently in particularly dusty conditions. If the element is allowed to become clogged with dust, engine performance and fuel consumption will suffer, and if the element is damaged or missing rapid internal wear may be expected. It is possible to work on the filter without any preliminary dismantling work, but it helps considerably if the legshield assembly (see Chapter 4) is removed.
2 After slackening the clamp which retains the intake hose to the carburettor, release the screws which retain the air filter cover and lift it away. The element is of the oiled foam type and is located over a supporting frame inside the filter casing.
3 Check the element carefully for holes or splits and renew it if damaged in any way. The element can be cleaned by washing in petrol. Once all dirt has been removed, dry the element and then soak it in clean engine oil. Squeeze out any excess to leave the element damp, but not dripping with oil. Refitting the element is a reversal of the removal operation, ensuring that the casing lid seals properly.

9 Oil pump: examination and renovation

1 The engine oil pump is unlikely to require attention, except during full engine overhauls, a desirable state of affairs since access to it requires the removal of the centrifugal clutch assembly. Once the pump assembly has been removed, it can be checked for wear as follows.
2 Remove the single screw which retains the pump cover to the body, and lift the cover away. A quick visual check of the cover, rotors and pump body will indicate whether the pump has been damaged by contaminated oil; any scoring will necessitate the renewal of the pump. If badly damaged, it may even be worth stripping the engine completely so that it can be cleaned out to prevent further damage.
3 The pump can be checked for wear by measuring the various clearances with feeler gauges and comparing them with the service limits given in the Specifications. Note that the rotor end clearance is checked by placing a straightedge across the pump gasket face and measuring between it and the pump rotors. If the pump is worn beyond the service limit in any dimension it should be renewed. Bear in mind that a worn pump can mean low oil pressure and an increased risk of general engine wear, so err on the side of caution.

8.3 Check air filter regularly. Renew element if torn or holed

9.1 Pump is driven off crankshaft. Remove clutch to gain access

Chapter 3 Ignition system

Contents

Specifications

Ignition system

Type	CDI (capacitor discharge ignition)
Advance method	Electronic
Static setting	7° BTDC at 1700 rpm
Advance setting	27° BTDC at 5000 rpm

Ignition coil

Winding resistances:	
Primary	1.6 ohm ± 10% at 20°C (68°F)
Secondary	6.6 ohm ± 10% at 20°C (68°F)

Source coil

Resistance - T50 model	312 - 468 ohm at 20°C (68°F)
Resistance - T80 model	240 ohm ± 10% at 20°C (68°F)
Lead colours	Black/red, White/black

Pickup coil

Resistance - T50 model	264 - 396 ohm at 20°C (68°F)
Resistance - T80 model	330 ohm ± 10% at 20°C (68°F)
Lead colours	White/red, Green/white

Spark plug

Type - T50 model	NGK CR6HS
Type - T80 model	NGK CR7HS
Electrode gap	0.6 - 0.7 mm (0.024 - 0.028 in)

1 General description

The ignition system is of the fully electronic CDI (capacitor discharge ignition) type. When the engine is running, electrical power is fed to the CDI unit, where a charge is stored in a capacitor which in this case acts like a small battery. When the CDI unit receives a trigger pulse, the capacitor discharges its stored current through the ignition coil primary windings. This in turn induces a high tension pulse in the ignition coil secondary windings, and this is applied to the spark plug. The high tension pulse jumps across the plug electrodes to earth, the resulting spark igniting the fuel/air mixture in the cylinder.

The trigger pulse is provided by a small magnetic pickup incorporated in the flywheel generator; a magnet in the rotor sweeps past a small pickup coil, generating a tiny signal voltage. This is sufficient to trigger the CDI unit, discharging the capacitor. The absence of mechanical parts in the system means that no maintenance is required as no adjustment is possible. This type of ignition system is well established and can be expected to be reliable over a long period of service. Where faults do occur, they tend towards failure of a single component in the system, and can be resolved by renewing the affected component.

1.1a General view of alternator stator assembly

1.1b Separate pickup coil is mounted on extension of stator

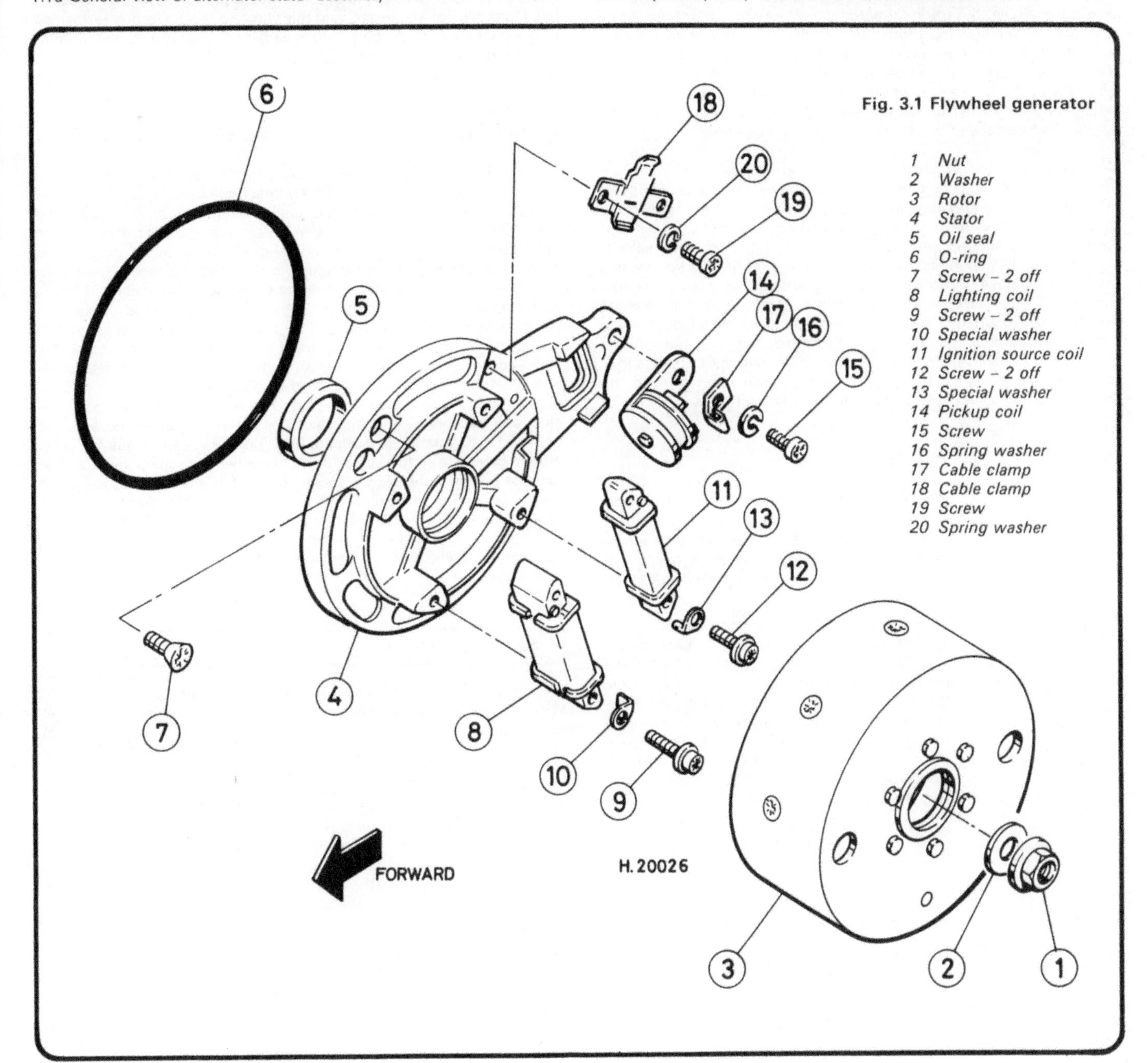

Fig. 3.1 Flywheel generator

1 *Nut*
2 *Washer*
3 *Rotor*
4 *Stator*
5 *Oil seal*
6 *O-ring*
7 *Screw – 2 off*
8 *Lighting coil*
9 *Screw – 2 off*
10 *Special washer*
11 *Ignition source coil*
12 *Screw – 2 off*
13 *Special washer*
14 *Pickup coil*
15 *Screw*
16 *Spring washer*
17 *Cable clamp*
18 *Cable clamp*
19 *Screw*
20 *Spring washer*

2 Checking the ignition system: general information

Test equipment

1 It is particularly important that any attempt to check a suspected fault in the ignition system is approached in a logical and methodical fashion. A haphazard attempt at resolving the fault is likely to prove time-consuming, confusing and often inconclusive. It should be noted that the range of tests which can be performed at home is distinctly limited, many of the more detailed tests requiring specialised test equipment unavailable to the owner by virtue of its cost. Whilst this means that in some cases it will be necessary to resort to an authorized Yamaha dealer for testing work or for confirmation of a suspected fault, it is quite reasonable for the owner to establish the general nature of the problem and to effect a cure in most instances.

2 A certain amount of specialist equipment will be needed to carry out the tests describe in this Chapter. For example, a stroboscopic timing lamp is needed to check the ignition timing, and a test meter will be required to make accurate resistance checks. For a full test of the CDI unit, a special test rig is necessary, and this will not be available for home use. In addition to the above, an extractor will be needed to draw off the generator rotor if access to the generator stator is necessary.

3 Of the above, the rotor extractor is well worth purchasing, and a stroboscopic timing lamp should be considered in view of its relatively low cost and general usefulness. A small multimeter (or 'pocket tester' as Yamaha describe it) is well worth thinking about, since it will prove generally useful for all sorts of electrical tests. A suitable meter can be ordered though Yamaha dealers or similar types can be purchased from car accessory shops or electronic component suppliers.

Defining the type of fault

4 Ignition faults can be divided into two main categories; partial or intermittent failure and complete failure. In the case of the former, the problem will often have developed gradually, commencing perhaps with unreliable starting, poor running or misfiring. Complete failure is less usual, and often indicates the failure of one of the ignition system components. Having decided which type of fault has occured, follow the sequence of checks as listed below, referring to the relevant Sections of this Chapter for details. Note that the checks are similar, whether the fault is partial or total. When dealing with the ignition system, treat it with respect, especially when switched on. A shock from the CDI system can be unpleasant or even dangerous to the owner, and may also cause damage to the CDI unit itself.

Check	Details
1) Spark plug	Always check the spark plug first; it is the single most likely cause of failure simply because it will eventually wear out, unlike the rest of the system. It is preferable to fit a new plug as a precautionary measure because a plug can sometimes prove faulty while appearing quite normal.
2) Spark	Fit a new plug into the plug cap and arrange it so that the body of the plug is in firm contact with a good earthing point, such as an unpainted part of the engine/transmission unit. Switch on the ignition and operate the kickstart lever while watching the plug electrodes. Under normal circumstances, a fat, regular bluish spark should be seen. A weak, irregular or yellow spark or no spark at all indicates a fault somewhere in the system.
3) Wiring and connectors	Using the wiring diagram at the back of the book, check through all of the ignition wiring, looking for damaged or broken leads. Check all connectors for looseness or corrosion. Remember that the earth connections are as important as the wiring connections.
4) Ignition coil, HT lead and plug cap	Check the coil connections as described above. Test the coil resistances as described in Section 7.
5) Pickup coil	Check the pickup coil resistance as described in Section 6. Renew the coil, if defective.
6) CDI unit	Check the CDI unit resistances, Section 8. If the fault persists, have the system checked by an authorized Yamaha dealer.
7) Ignition timing	The ignition timing should be checked in cases where an intermittent fault or poor running cannot be traced to another cause. Inaccurate timing means renewal of the CDI unit.

3 Spark plug: maintenance and renewal

1 In normal circumstances the standard grade spark plug should be used, but if the machine is used in constantly low temperatures, or is ridden hard in hot conditions, a change of grade may be required. If problems are experienced, it is advisable to consult a Yamaha dealer for advice on any proposed change. When fitting a new plug, always set the electrode gap to the prescribed 0.6 – 0.7 mm (0.024 – 0.028 in). It is quite permissible to clean and readjust a used plug, which can then be re-used. Many owners prefer to renew the plug in preference to cleaning. Plugs are not expensive, and this choice of action removes any risk of being let down by a 'tired' plug at a later date. If cleaning and refitting is preferred, this is described below.

2 Before cleaning or discarding the plug, refer to the colour section which accompanies this Chapter. This gives useful information on diagnosing general engine condition and mixture strength based on the appearance of the plug electrodes, and can be a helpful diagnostic aid.

3 The plug electrodes can be cleaned by carefully scraping away the accumulated carbon deposits. This is best done using a combination of a small knife blade and assorted small files and abrasive paper. Take great care to avoid straining the centre electrode, or the ceramic insulator nose may be chipped or broken. Alternatively, the plug can be cleaned by abrasive blasting, either in a full-size garage machine or with one of the small DIY models now available. Once clean, file the opposing faces of the electrodes flat, using a small, fine file. A magneto file or even a nail file can be used for this purpose. Whichever method is chosen, make sure that every trace of abrasive and loose carbon is removed before the plug is refitted. If this is not done, the debris will enter the engine and cause damage or rapid wear.

4 Whether a cleaned or new plug is to be fitted, always check the electrode gap before it is installed. Use feeler gauges to measure the gap, and if adjustment is required, bend the outer, earth electrode only. **Never** bend the centre electrode or the ceramic insulator nose will be damaged. Before the plug is fitted, apply a fine coat of PBC grease to the threads. This will help prevent thread wear and damage. Fit the plug finger tight, then tighten it a further 1/4 turn only, to ensure a gas-tight seal. Beware of overtightening, and always use a plug spanner or socket of the correct size.

4 Spark plug: check

1 This is an essential first step in tracing any suspected ignition fault. With the ignition switched off, remove the spark plug from the cylinder head and refit it in the plug cap. Place the metal body of the plug in firm contact with a good earth point, such as the unpainted metal of the engine/transmission unit, and in a position where the plug electrodes can be viewed while the kickstart is operated.

2 Switch on the ignition and operate the kickstart lever while watching the plug electrodes. If all is well, a regular, fat bluish spark should be seen jumping across the electrode gap. A thin, yellowish or irregular spark is indicative of a fault. If the spark looks weak, try fitting a new plug, having set the electrode gap correctly. If this fails to improve the spark, check the ignition coil resistances as described below.

3 Where no spark at all is evident, check for a break in the ignition system wiring or for a fault in the ignition switch. Try turning the switch on and off a few times if it is suspected that the switch contacts are dirty or corroded. If the fault persists, it will be necessary to check through the ignition system components as described in the following Sections until the cause of the problem is isolated.

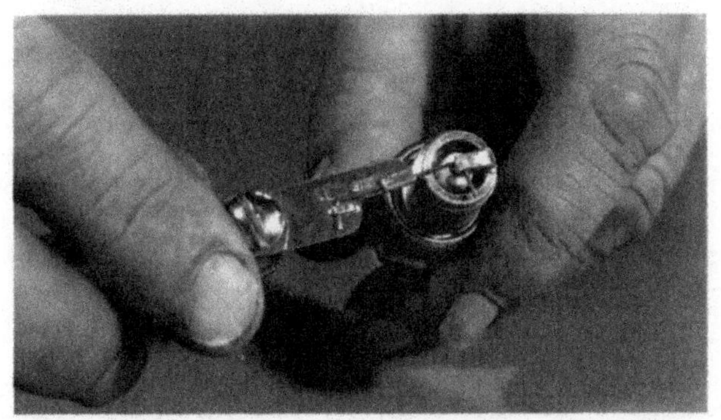

Spark plug maintenance: Checking plug gap with feeler gauges

Altering the plug gap. Note use of correct tool

Spark plug conditions: A brown, tan or grey firing end is indicative of correct engine running conditions and the selection of the appropriate heat rating plug

White deposits have accumulated from excessive amounts of oil in the combustion chamber or through the use of low quality oil. Remove deposits or a hot spot may form

Black sooty deposits indicate an over-rich fuel/air mixture, or a malfunctioning ignition system. If no improvement is obtained, try one grade hotter plug

Wet, oily carbon deposits form an electrical leakage path along the insulator nose, resulting in a misfire. The cause may be a badly worn engine or a malfunctioning ignition system

A blistered white insulator or melted electrode indicates over-advanced ignition timing or a malfunctioning cooling system. If correction does not prove effective, try a colder grade plug

A worn spark plug not only wastes fuel but also overloads the whole ignition system because the increased gap requires higher voltage to initiate the spark. This condition can also affect air pollution

5 Checking the ignition system: tracing wiring and switch faults

1 A broken or badly connected wire, or dirty or contaminated switch contacts can often be traced as the cause of an ignition or electrical fault. Using the wiring diagram at the back of this manual, trace through the ignition wiring and check each connector for security, corrosion and the presence of water, paying particular attention to the CDI unit connector. Look for chafed or broken wires and renew or repair these as necessary. Check the ignition switch for water contamination (see Chapter 6).
2 A quick cure for water in switch or wiring connectors is to spray them with a silicone-based maintenance aerosol, such as WD40. A better long-term measure is to dismantle all connectors, pack them with silicone grease, and then reassemble them. Done properly, the machine will run virtually submerged in water, and there should be no further trouble of this type.

6 Alternator coils: testing

1 The ignition source coil is mounted on the flywheel generator stator plate and provides power for the system. Mounted externally to the rotor is the pickup, or pulser, coil which provides the trigger signal which controls the ignition spark. If it is suspected that a fault lies in either of the above coils this can be checked by using a multimeter to test the coil resistances. Start by removing the left-hand side panel to gain access to the various wiring connectors.
2 Trace the flywheel generator wiring back to the connectors, and separate them. Using a multimeter set to the appropriate resistance range measure the resistance of the source coil and pickup coil windings. Connect the meter probes between the black/red and white/black wires and note the reading for the source coil, then connect the probes between the white/red and green/white wires and note the reading for the pickup coil.
3 If the test shows a reading of zero ohms, the coil insulation has broken down, allowing the coil to short to earth. If on the other hand, a reading of infinite resistance is shown, the coil windings have broken. Either fault will require the renewal of the affected coil but before ordering a new item it is worth having your findings confirmed by an expert.
4 The only real alternatives to renewal are to consult an auto-electrical specialist, who may be able to rewind the faulty coil, or to consider buying a second-hand item from a motorcycle breaker. Removal and refitting of the alternator is described in the relevant Sections of Chapter 1.

7 Ignition coil: testing

1 The ignition coil is mounted on the frame tube, below the legshield moulding. To gain access to the coil it will first be necessary to remove the legshield, see Chapter 4. Disconnect the HT lead and low tension lead. If it is suspected that a fault lies in the ignition coil this can be checked by using a multimeter to test the coil resistances.
2 Using a multimeter set on the resistance x1 range, check the resistance of the primary windings by connecting one probe to the orange lead terminal on the coil and the second to earth (ground). A reading of 1.6 ohms ± 10% should be indicated.
3 Next, measure the secondary resistance by connecting one probe to earth (ground) and the other to the HT lead. A reading of 6.6 k ohms ± 10% should be shown.
4 If the test gave incorrect resistance readings, the coil is suspect. Note that the resistance tests are not exhaustive: for a more accurate diagnosis, have the coil tested on a proper coil testing machine.

8 CDI unit: testing

1 In the event of a suspected fault in the CDI unit, a certain amount of testing is possible at home, given access to a suitable multimeter. To gain access to the CDI unit, remove the left-hand side panel. The unit is mounted next to the finned alloy regulator/rectifier unit. Trace the wiring back to the connector below the unit and unplug it, taking care to pull on the connector body only, not on the wiring. Before performing the resistance tests, check the wiring connector for water contamination or corrosion, these being the most likely faults.
2 Check the various resistances as shown in the accompanying table. The CDI unit terminals are identified in the line drawing which accompanies the table.
3 It must be stressed that the above test gives only a rough indication of the condition of the unit. A full test can only be conducted using a special test rig which may be available at a local Yamaha dealer. Failing this, the only alternative is to check the unit by substitution.

7.1 Ignition coil is attached to bracket below frame main tube

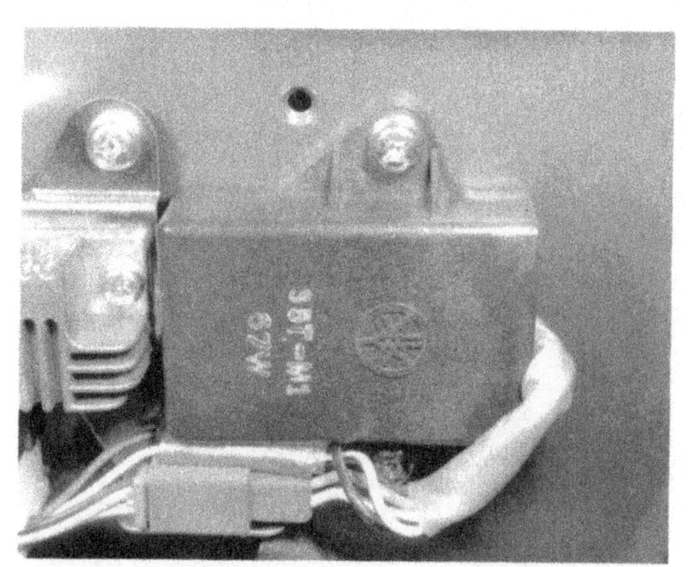

8.1 CDI unit is housed behind left-hand side panel

		TESTERS BLACK LEAD WIRE					
		○	W/B	W/R	G/W	B/R	B
TESTERS RED LEAD WIRE	○		∞	∞	∞	∞	∞
	W/B	∞		∞	∞	∞	∞
	W/R	∞	13 kΩ ~ 15 kΩ		50 kΩ ~ 70 kΩ	13 kΩ ~ 15 kΩ	○
	G/W	∞	∞	∞			∞
	B/R	∞	50 kΩ ~ 100 kΩ	3 kΩ ~ 7 kΩ	30 kΩ ~ 50 kΩ		3 kΩ ~ 7 kΩ
	B	∞	13 kΩ ~ 15 kΩ	○	50 kΩ ~ 70 kΩ	3 kΩ ~ 7 kΩ	

H.20029

Fig. 3.2 CDI unit test table

B *Black*
B/R *Black and red*
G/W *Green and white*
W/R *White and red*
W/B *White and black*

9 Ignition timing: checking

1 The ignition timing check need only be performed if there is some reason to suspect its accuracy. It should be noted that no adjustment is possible, and that the only reason for a timing error is a fault in the CDI unit. Yamaha do not normally recommend a timing check as being necessary. The test requires the use of a stroboscopic timing lamp ('strobe'), preferably of the xenon tube type, rather than the cheaper and less accurate neon variety. In the absence of this equipment, have the test carried out by an authorized Yamaha dealer.

2 Remove the legshield assembly (see Chapter 4) and gearchange lever, then remove the crankcase left-hand side cover to reveal the rotor edge. Identify the timing 'F' mark on the rotor, and the fixed index mark on the crankcase, above the rotor. Connect the strobe, following the manufacturer's instructions, start the engine and direct the beam of the strobe at the fixed index mark. The timing is correct if the 'U' mark aligns with the index mark. If the timing seems incorrect, check the CDI unit and the source and pickup coils as described above.

Chapter 4 Frame and suspension

Contents

Specifications

Frame

Type	Welded tubular and pressed steel spine

Front suspension

Type	Leading link
Front suspension units	Hydraulically-damped coil spring
Front suspension travel	40 mm (1.6 in)
Front wheel travel	80 mm (3.2 in)
Spring free length – early T80 original units	168 mm (6.614 in)
Spring free length – all other models and replacement units for early T80	178 mm (7.008 in)

Rear suspension

Type	Swinging arm
Suspension units	Oil-damped coil spring telescopic
Rear suspension travel	60 mm (2.4 in)
Rear wheel travel	70 mm (2.8 in)
Spring free length	213 mm (8.386 in)

Torque wrench settings

Component	kgf m	lbf ft
Front wheel spindle	3.9	28.0
Front suspension link pivot bolts	1.8	13.0
Front suspension unit mounting bolts:		
Lower	2.0	14.0
Upper	2.0	14.0
Handlebar nacelle retaining nuts	1.6	11.0
Handlebar bracket bolts	1.6	11.0
Handle plate to handle crown bolts	2.4	17.0
Handle crown bolts to steering column	1.6	11.0
Steering stem locknut	3.0	22.0
Swinging arm pivot	5.6	40.0
Rear suspension unit mountings:		
Upper	3.0	22.0
Lower	2.3	17.0
Rear brake torque arm	1.5	11.0
Final drive bevel gearbox:		
Gearbox side cover bolts	1.0	7.0
Gearbox to swinging arm bolts	2.6	18.8
Drive pinion bearing retainer	6.0	43.0

1 General description

The Yamaha T50 and T80 models employ a welded steel 'step-through' type frame comprising a large diameter tubular front section, welded to a pressed steel main section. The front frame loop runs low between the steering head and the pressed steel frame rear section, and is normally covered by an injection-moulded plastic legshield. The engine/transmission unit is mounted below it. The rear section forms a hollow shell which is used to house the fuel tank and some electrical components, and also to provide mounting points for the rear suspension.

Front suspension takes the form of a single long steering column, at the lower end of which is welded a U-shaped fork, this being fabricated from sheet steel pressings. From each end of the fork, a short leading link runs forward, the ends of which incorporate bosses for the front wheel spindle. The rear of each link is connected to the fork by way of small pivot pins, whilst telescopic suspension units are connected at the centre of each link, these running inside the fork pressings.

Rear suspension is by a conventional pivoted rear fork, or swinging arm, arrangement. The cast alloy swinging arm is attached to the frame by a pivot shaft, the assembly being supported on a pair of oil-damped telescopic suspension units.

Unusually, for a machine of this type, shaft final drive is employed. Power is taken from the gearbox output shaft to a middle gear case at the rear of the crankcase casting. Here drive is turned through right angles by way of a pair of bevel gears. From the middle gear case, power is fed through a universal joint to allow swinging arm movement, and then along a steel shaft housed in the left-hand side of the swinging arm casting to the rear bevel housing. At the rear of the shaft, drive turns through 90° once more through a second bevel gear set, and is applied to the rear wheel.

2 Front suspension: dismantling, examination and reassembly

1 It is not necessary to remove the front fork assembly from the machine if attention to the front suspension is required, the suspension components being quite distinct from the steering assembly. Start by placing the machine on its centre stand on level ground, then raise the front wheel clear of the ground by placing wooden blocks or a jack below the crankcase. Remove the wire circlip which secures the speedometer cable to the brake backplate. Slacken and remove the front brake cable adjuster nut and disengage the cable from the operating arm. Refit the trunnion and nut on the cable for safe keeping.

2 Straighten and remove the split pin from the wheel spindle nut. Slacken and remove the nut, then support the weight of the wheel while the wheel spindle is displaced and removed. The wheel can now be manoeuvred clear of the forks and placed to one side.

3 Unscrew the suspension unit upper mounting nuts. These are located on the inner face of the fork. Slacken and remove the suspension link pivot bolt locknuts at the lower end of the forks, then remove the pivot bolts. The links, together with the suspension units, can be withdrawn for further dismantling and examination.

4 Clean each suspension assembly carefully to remove the accumulated road dirt. Remove the suspension unit lower mounting locknuts, then unscrew the cross-head lower mounting bolts and separate the units from the links.

5 From the suspension unit lower mountings and the link pivots, remove the dust caps, the felt dust seals and the collars. Check the various parts for wear or corrosion. Light corrosion can be cleaned up using fine abrasive paper, provided that this does not result in a sloppy fit between the collars and the bushes.

6 Note that if the suspension link bushes are in need of renewal, removal is unlikely to be easy. These bushes can be extremely difficult to dislodge, particularly where corrosion has occurred between the bush and the suspension link. It will be a lot quicker and easier if they can be pressed out. Failing that, soak them in a releasing fluid like Plus Gas, then make up a drawbolt arrangement such as that shown in the accompanying line drawing.

7 Using the drawbolt or press, extract the worn bush and fit the new one, ensuring that it is located centrally in its boss. If removal proves unusually difficult, try heating the link in a domestic oven or by using a hot air gun of the type used for stripping paint. Ensure that all traces of releasing fluid have evaporated before subjecting the link to heat. The heat will help to break down the corrosion between the two components.

8 Before assembly commences, check the suspension units for signs of leakage or other damage. Little can be done by way of repair; if the units are defective, they must be renewed. Always fit replacement units as a pair, never singly.

9 Grease and assemble the bushes and collars. Soak the felt dust seals in engine oil, then position them in their recesses on the link. Fit the dust caps, retaining them in position by peening the sides of each one in two or three places using a punch or cold chisel. Refit the suspension unit lower mounting bolts, tightening them securely. Refit the mounting bolt locknuts, tightening them to 2.0 kgf m (14.0 lbf ft).

10 Offer up each assembly in turn, fitting the suspension unit upper mounting nuts and the link pivot bolts loosely. Tighten the upper mounting nuts to 2.0 kgf m (14.0 lbf ft), then tighten the link pivot bolt and nut to 1.8 kgf m (13.0 lbf ft).

11 Check that the front brake backplate is fitted correctly and that the speedometer drive engages correctly over the hub slots. Offer up the front wheel, aligning the slot on the brake backplate with the boss on the left-hand suspension link. Fit the wheel spindle (from the right-hand side), then fit and tighten the wheel spindle nut to 3.9 kgf m (28.0 lbf ft). Reconnect the speedometer cable and the front brake cable. Set the brake cable adjuster to give 5 – 8 mm (0.2 – 0.3 in) free play, measured between the lever stock and blade.

2.3a Remove the suspension unit upper mounting nuts and displace retaining bolts

2.3b Remove pivot bolt nuts, displacing bolts to free suspension links

2.3c Remove bolts and cup washers to free mudguard, if required

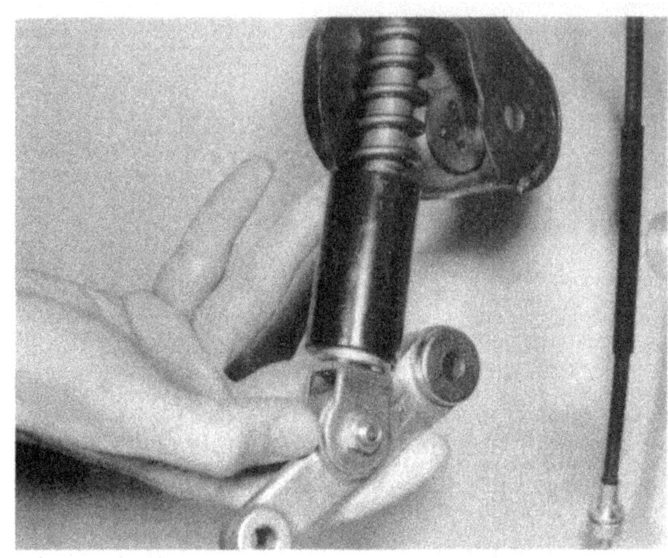

2.3d Suspension links and unit can now be withdrawn from fork legs

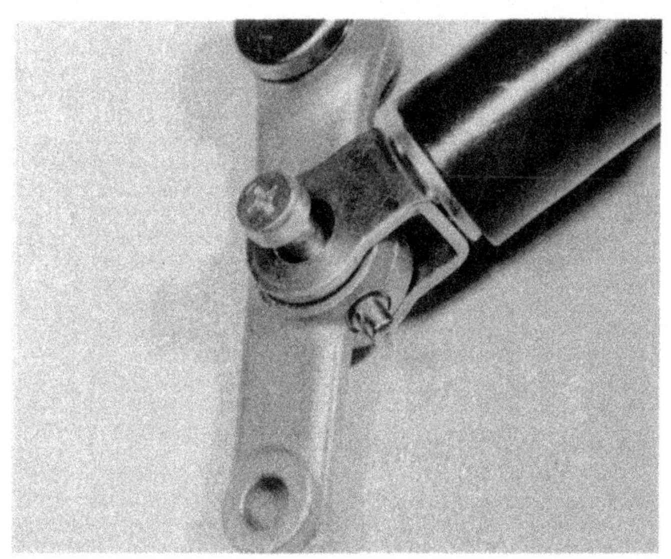

2.4 Remove locknuts and unscrew lower mounting screw to free links from suspension units

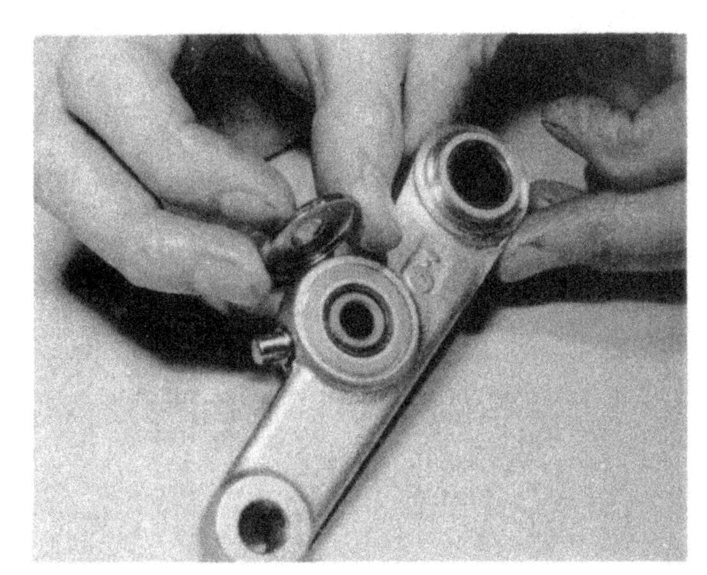

2.5a Displace and remove dust caps and collars

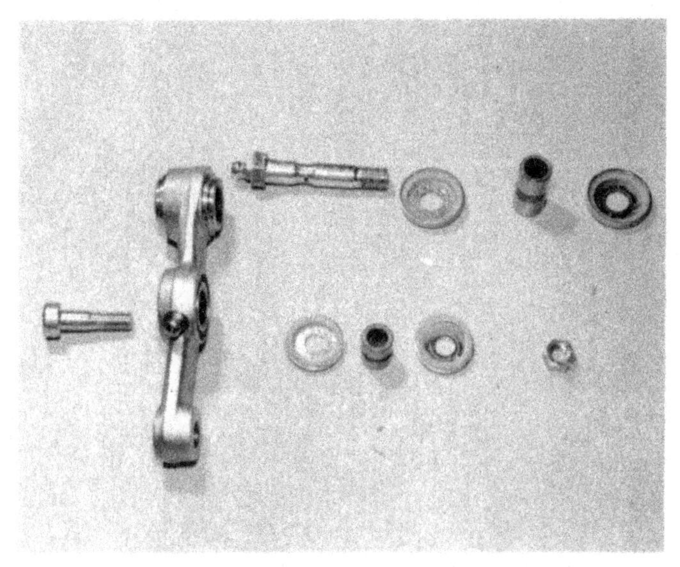

2.5b The suspension link components laid out for examination

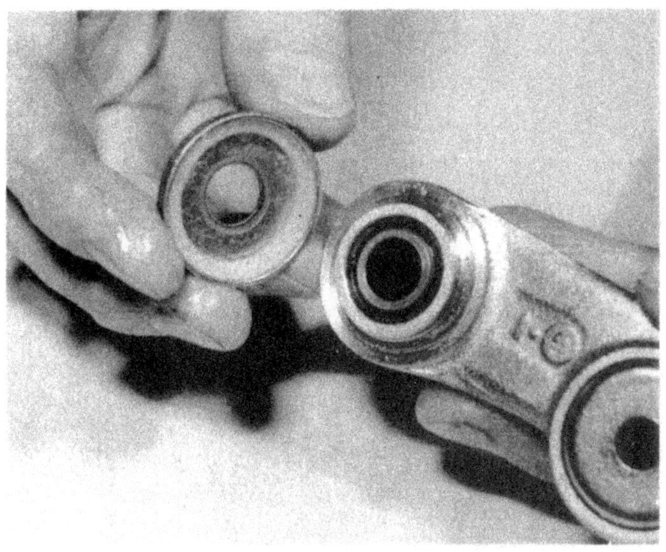

2.9 Grease collars and oil felt seals during assembly

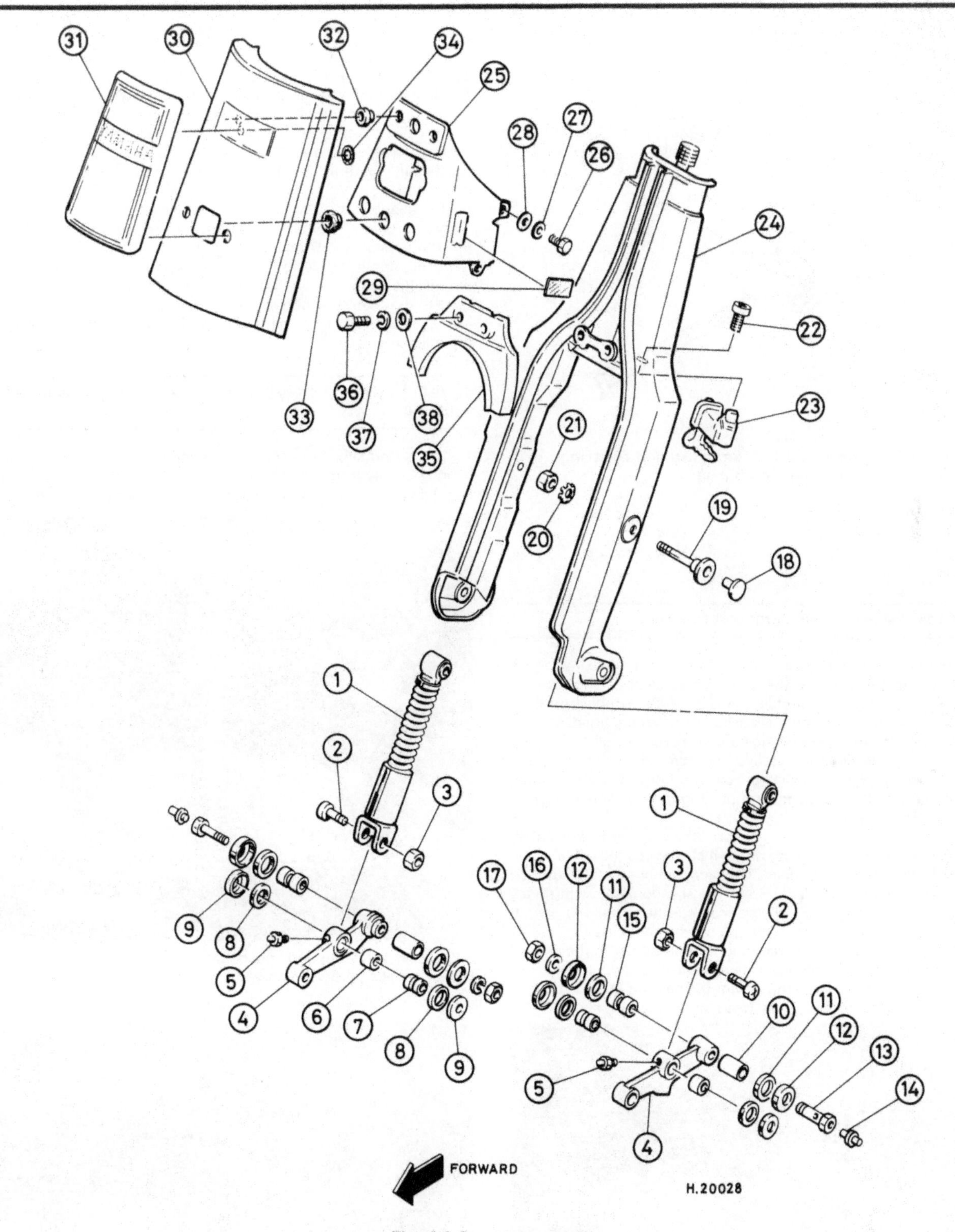

Fig. 4.1 Front suspension

1 Suspension unit – 2 off
2 Screw – 2 off
3 Nut – 2 off
4 Suspension link – 2 off
5 Grease nipple – 2 off
6 Bush – 2 off
7 Collar – 2 off
8 Felt seal – 4 off
9 Dust cap – 4 off
10 Bush – 2 off
11 Felt seal – 4 off
12 Dust cap – 4 off
13 Bolt – 2 off
14 Grease nipple – 2 off
15 Collar – 2 off
16 Spring washer – 2 off
17 Nut – 2 off
18 Plug – 2 off
19 Bolt – 2 off
20 Lock washer – 2 off
21 Nut – 2 off
22 Screw
23 Steering lock
24 Steering column
25 Steering column bracket
26 Bolt – 4 off
27 Spring washer – 4 off
28 Washer – 4 off
29 Damping rubber – 4 off
30 Steering column front cover
31 Emblem
32 Grommet – 2 off
33 Grommet
34 Fastener – 2 off
35 Lower cover
36 Bolt – 2 off
37 Spring washer – 2 off
38 Washer – 2 off

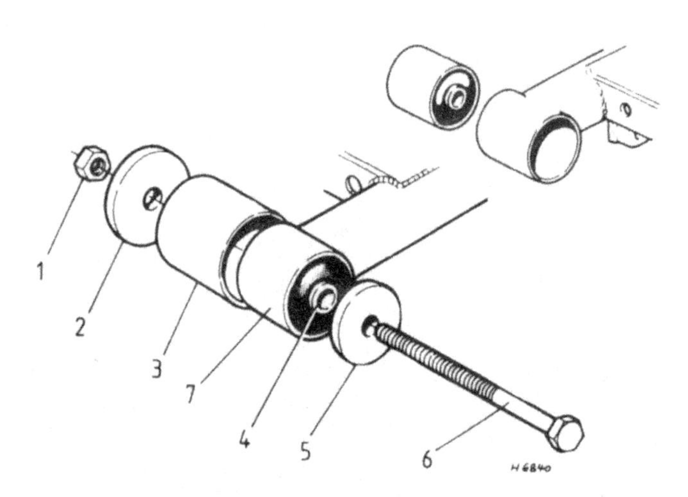

Fig. 4.2 Drawbolt tool for removing and refitting suspension bushes

1 *Nut*
2 *Thick washer*
3 *Tube*
4 *Bush*
5 *Thick washer*
6 *Bolt*
7 *Bush housing*

3 Handlebar assembly: removal and refitting

1 Place the machine on its centre stand and arrange blocks or a jack beneath the machine to raise the front wheel clear of the ground. Remove the six screws which retain the legshield assembly and lift it away, taking care not to scratch the plastic finish. Where fitted, remove the front carrier. Remove the steering column cover by pulling it off its rubber mounts. Slacken and remove the two headlamp securing screws and disengage the headlamp from the front of the handlebar nacelle. Disconnect the headlamp and parking lamp wiring and place the headlamp to one side.
2 Wrap some PVC tape around the top of the speedometer cable, then unscrew the knurled retaining ring which secures the cable to the underside of the speedometer head. The tape will prevent the knurled ring from slipping down the cable. Trace the speedometer wiring back to the connectors and separate them. Separate the remaining switch harness and wiring connectors inside the headlamp recess.
3 Disconnect the choke, front brake and throttle cables. In the case of the throttle cable, separate the right-hand handlebar switch halves to allow the cable to be freed from the twistgrip. The front brake cable can be disconnected after it has been freed at the wheel end to give maximum free play, whilst the choke cable should be freed at the carburettor for the same reason.
4 Remove, from the underside of the assembly, the two nuts which secure the lower half of the nacelle to the upper half. The upper half of the nacelle, together with the handlebar assembly, can now be lifted away and placed to one side. Unscrew the three bolts which retain the handlebar mounting bracket and remove it, followed by the nacelle lower half.
5 Reassembly is a reversal of the dismantling sequence, noting the following points. Tighten the handlebar bracket bolts and the nacelle retaining nuts to 1.6 kgf m (11.0 lbf ft). When fitting the throttle twistgrip, apply grease to the handlebar end on which it runs, and to the throttle cable groove in the pulley section. Remember to check and adjust the throttle cable and front brake cable (see Routine Maintenance) and to check the operation of all controls before riding the machine.

4 Front fork and steering head assembly: removal and refitting

1 It will normally be necessary to remove the front fork and steering head assembly from the machine only in the event of the steering head bearings requiring attention, or if the fork itself has been damaged as

3.2 Access to wiring and speedometer cable is improved with top half of nacelle partially removed

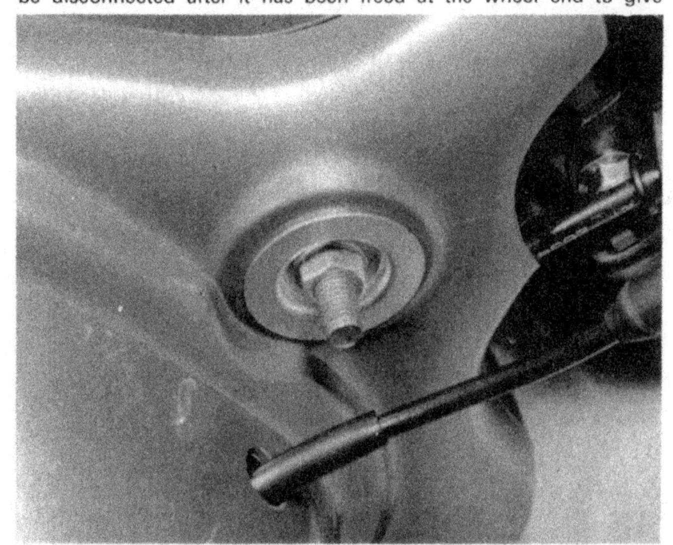

3.4a Release nuts and cup washers to free nacelle top half

3.4b Remove nacelle and handlebar as an assembly

3.4c Handlebar bracket is secured by three bolts

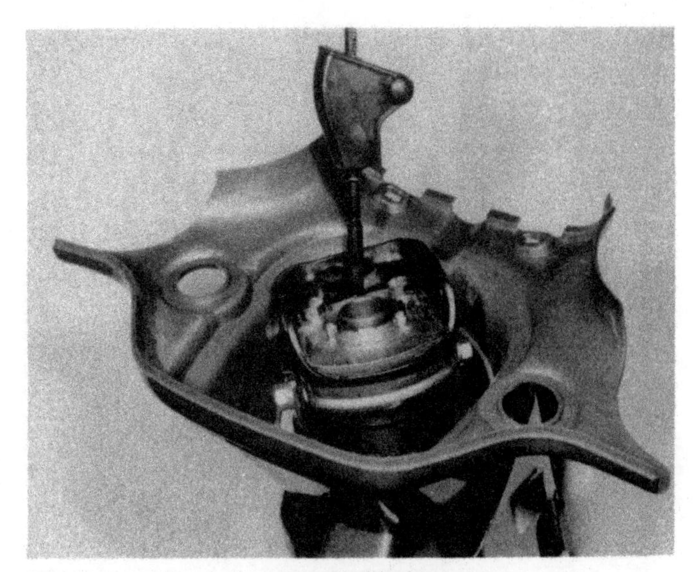

3.4d Nacelle lower half can now be lifted away

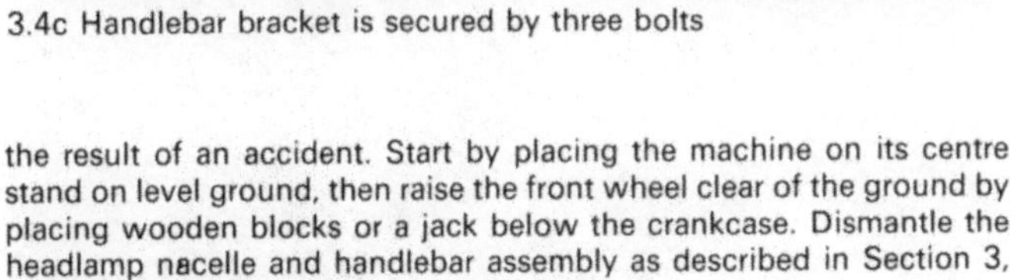

the result of an accident. Start by placing the machine on its centre stand on level ground, then raise the front wheel clear of the ground by placing wooden blocks or a jack below the crankcase. Dismantle the headlamp nacelle and handlebar assembly as described in Section 3, then remove the front wheel as follows:

2 Prise out the wire circlip which secures the speedometer cable to the brake backplate. Slacken and remove the front brake cable adjuster nut and disengage the cable from the operating arm. Refit the trunnion and nut on the cable for safekeeping. Straighten and remove the split pin from the wheel spindle nut. Slacken and remove the nut, then support the weight of the wheel while the wheel spindle is displaced and removed. The wheel can now be manoeuvred clear of the forks and placed to one side.

3 Slacken and remove the two bolts and the tubular spacers which pass down through the top plate assembly, leaving access clear to the steering stem nut. To remove this nut, a C-spanner will be required. Slacken and remove the nut, then lift away the top plate and the rubber damper. The handle plate and the second rubber damper can now be lifted away. Unscrew the bolt on each side of the steering column to permit handle crown removal.

4 Support the fork assembly, then slacken and remove the bearing adjuster nut. The fork assembly can now be lowered clear of the steering head and removed, taking care not to lose any of the steering head balls if they drop free. Remember to clean and examine the bearings and races prior to reassembly.

5 Commence reassembly by coating the steering head bearing balls in heavy grease and using this to stick them in place on the steering head bearing cups. There should be 22 balls in each race; this allows a small amount of free space between the balls to allow free movement of the races. Carefully refit the steering stem through the bearing races, taking care not to dislodge the balls. Place the upper cone in position, then fit the adjuster nut finger-tight to secure the assembly.

6 Tighten the adjuster nut until it stops, then back it off by about $^1/_8$ of a turn. The adjuster should be set so that the steering moves freely, but with no detectable free play in the bearings. Continue reassembly in the reverse of the dismantling sequence, noting that the handle crown to steering column bolts should be tightened to 1.6 kgf m (11 lbf ft) and the handle plate to handle crown bolts, which pass down through the top plate and damper, should be tightened to 2.4 kgf m (17 lbf ft). Tighten the steering stem locknut to 3.0 kgf m (22.0 lbf ft). Refit the handlebar, controls and nacelle as described in Section 3.

7 Check that the front brake backplate is fitted correctly and that the speedometer drive engages correctly over the hub slots. Offer up the front wheel, aligning the slot in the brake backplate with the boss on the left-hand suspension link. Fit the wheel spindle (from the right-hand side), then fit and tighten the wheel spindle nut to 3.9 kgf m (28 lbf ft). Reconnect the speedometer cable and the front brake cable. Set the brake cable adjuster to give 5 – 8 mm (0.2 – 0.3 in) free play, measure between the lever stock and blade.

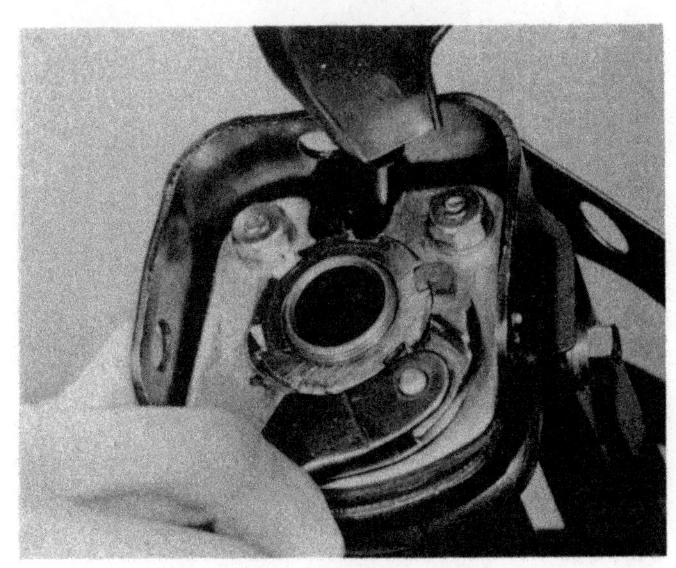

4.3a Slacken and remove the steering stem locknut

4.3b Remove the two bolts and lift away the top plate ...

4.3c ... followed by the rubber damper and handle plate ...

4.3d ... and the spacers and second rubber damper

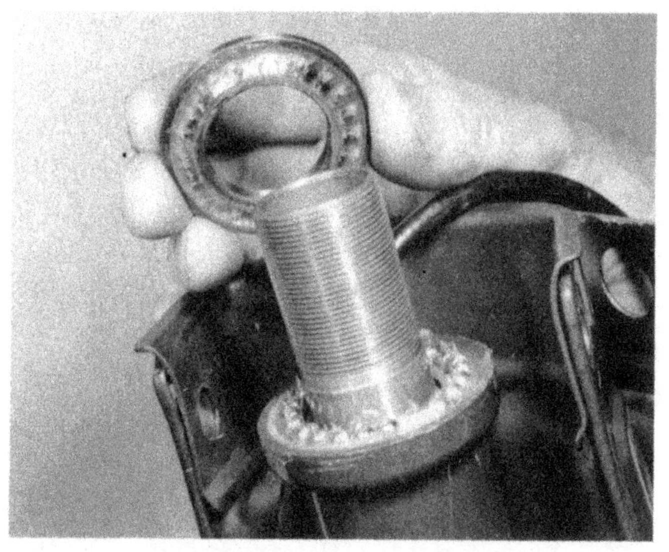

4.5a Coat steering head bearing balls with grease to hold them in place, then refit the fork assembly and top cup

4.5b Refit the shroud, and set the bearing adjuster nut

4.5c Continue reassembly of the steering stem components ...

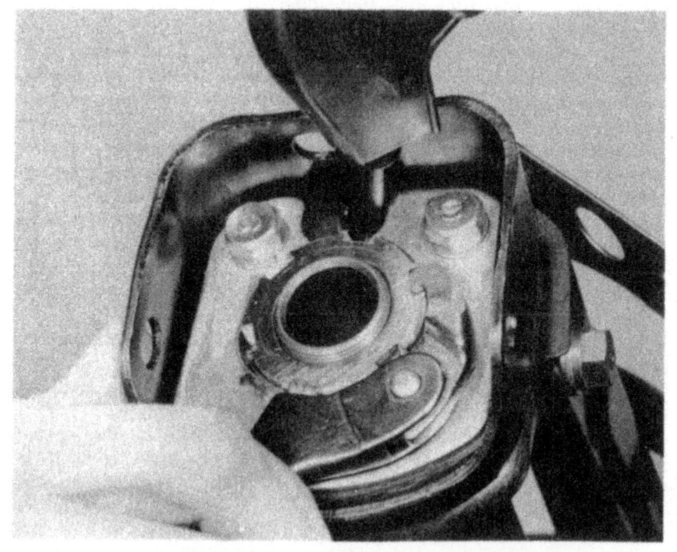

4.5d ... tightening the locknut to preserve bearing adjustment

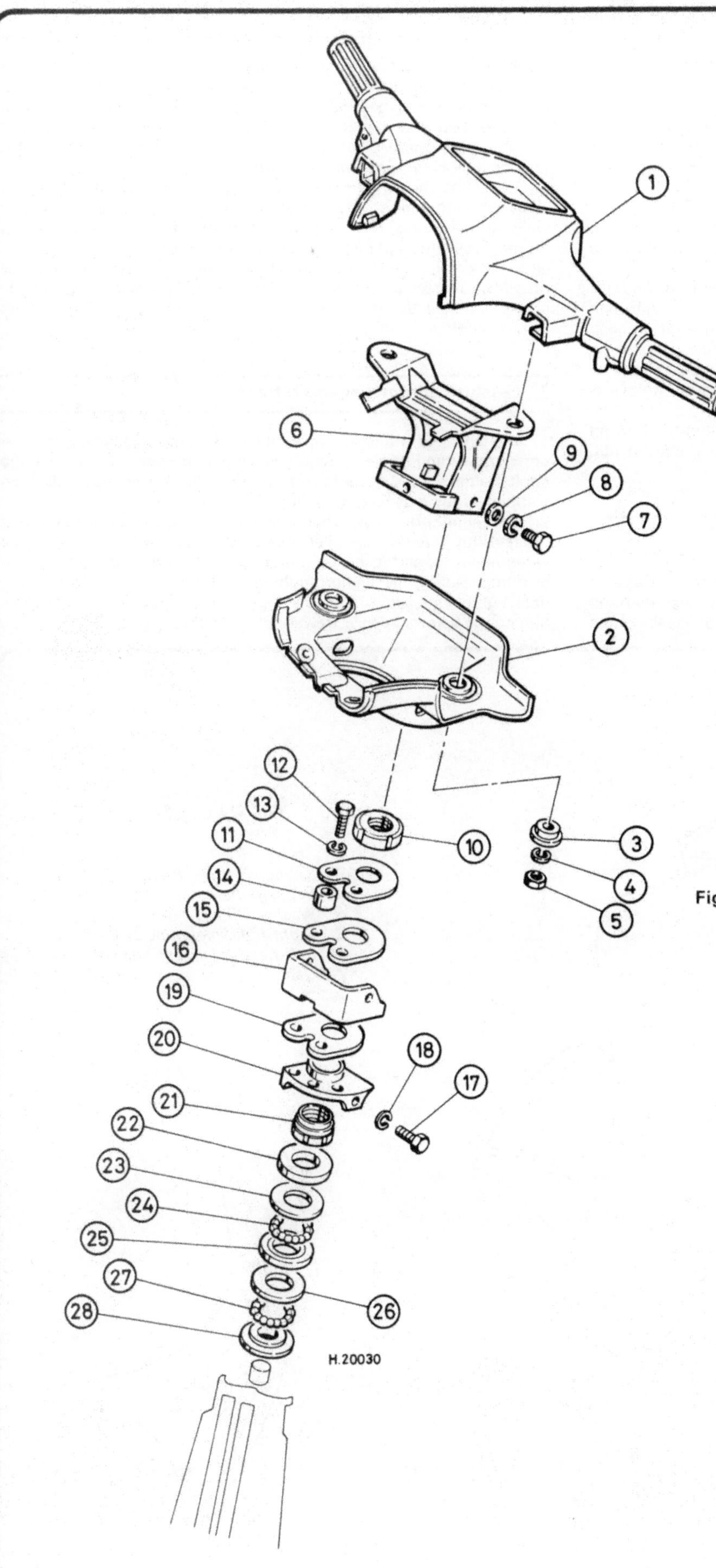

Fig. 4.3 Handlebar and steering head assembly

1 *Handlebar/upper headlamp nacelle*
2 *Lower headlamp nacelle*
3 *Cup washer – 2 off*
4 *Spring washer – 2 off*
5 *Nut – 2 off*
6 *Handlebar bracket*
7 *Bolt – 3 off*
8 *Spring washer – 3 off*
9 *Washer – 3 off*
10 *Locknut*
11 *Top plate*
12 *Bolt – 2 off*
13 *Spring washer – 2 off*
14 *Spacer – 2 off*
15 *Rubber damper*
16 *Handle plate*
17 *Bolt – 2 off*
18 *Spring washer – 2 off*
19 *Rubber damper*
20 *Handle crown*
21 *Bearing adjuster nut*
22 *Dust cover*
23 *Upper bearing cone*
24 *Upper bearing balls*
25 *Upper bearing cup*
26 *Lower bearing cup*
27 *Lower bearing balls*
28 *Lower bearing cone*

5 Steering head bearings: examination and renovation

1 The steering head bearing balls, cups and cones should be examined for wear if there are signs of roughness when the steering is turned from lock to lock. To gain access to the bearings it is necessary to dismantle the steering head assembly as described in Section 4.
2 Examine the balls and the cups and cones for signs of pitting or indentations. If found, it is preferable to renew the entire assembly, in view of the amount of dismantling work required to gain access to them.
3 The bottom cone will remain attached to the steering stem, and this can be removed by tapping an old screwdriver underneath it and then working it up and off the steering stem, using a screwdriver on each side of the cone.
4 The bearing cups are pressed into the steering column and may be removed using a long drift. The new races can be tapped into place, making sure that they enter their bores squarely.
5 When reassembling the steering head, make sure that the bearing adjustment is set correctly; it is easy to overtighten the adjuster nut, and this will cause rapid wear of the balls and the races.

6 Frame: examination and renovation

1 The frame is unlikely to require very much attention unless it becomes damaged in an accident, in which case specialist attention will be required to assess the viability or otherwise of repairing the machine. If the frame becomes badly bent, it is preferable to fit a replacement. Depending upon the nature of the damage and the age of the machine, it may be worth trying to purchase a second machine of the same year and model with, perhaps, a damaged engine. This is likely to work out less costly than buying a new frame, and will also provide a stock of used spares for future use.
2 The frame should be checked periodically for signs of cracking in areas of high stress. These should be repaired professionally by welding or brazing. The construction of this type of frame makes it prone to the effects of rusting. Keep an eye on likely areas of corrosion, and rub down, prime and repaint any damaged areas as soon as possible to prevent it from spreading. Yamaha dealers can supply colour-matched paints in aerosol cans for touching up rust or stone chip damage.

7 Legshields: removal and refitting

1 The legshield is retained by a total of six cross-headed screws, the upper right-hand screw doubling as a mounting point for the luggage hook. Remove the six screws, working from the bottom upwards, then manoeuvre the legshield moulding clear of the frame.
2 Care should be taken when handling the legshield moulding; it is resilient but is easily scratched. Place it on soft cloth in a safe place when removed, and take care not to score the surface during removal or fitting. During installation, tighten all fasteners finger-tight before final tightening. Check that the legshield is aligned correctly, then tighten the screws evenly. Beware of overtightening.

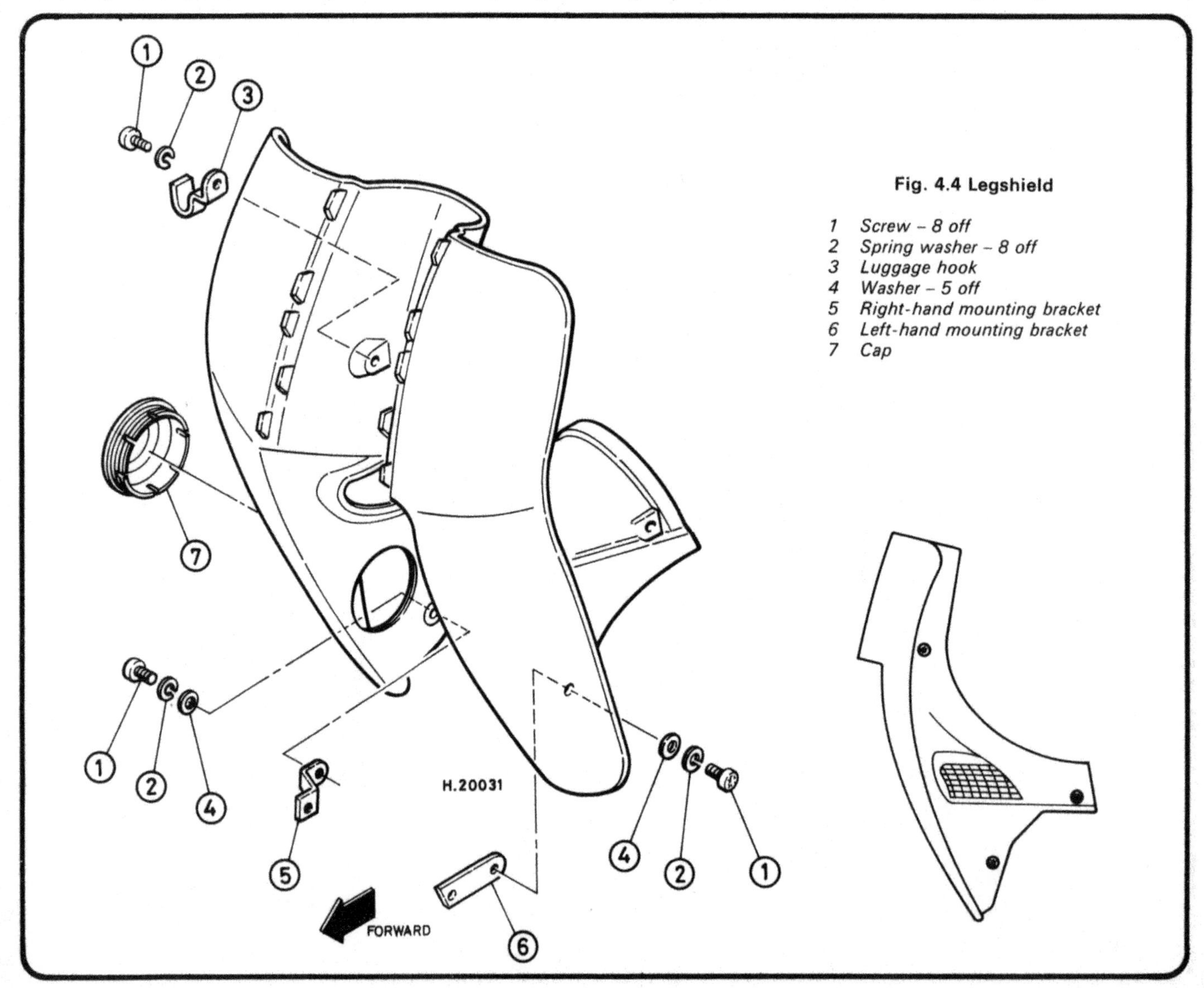

Fig. 4.4 Legshield

1 *Screw – 8 off*
2 *Spring washer – 8 off*
3 *Luggage hook*
4 *Washer – 5 off*
5 *Right-hand mounting bracket*
6 *Left-hand mounting bracket*
7 *Cap*

8 Swinging arm: examination and renovation

1 The swinging arm is supported on a pivot shaft which screws into the driveshaft casing. The shaft is carried on bushes pressed into the frame cross-tube. To check for wear, place the machine on its centre stand and attempt to push and pull the swinging arm sideways. A very small amount of movement is acceptable, but if there is excessive play, the assembly should be dismantled and overhauled as described below.

2 Remove the rear brake adjuster nut and the brake torque arm nut from the brake backplate. Disengage the brake operating cable from the arm, refitting the trunnion and nut on the end of the rod for safekeeping. Free the torque arm from the brake backplate.

3 Straighten and remove the split pin from the rear wheel spindle nut. Slacken and remove the wheel spindle nut and withdraw the wheel spindle. Displace the spacer on the right-hand side of the wheel, then pull the wheel away from the cush drive hub and manoeuvre it clear of the swinging arm.

4 Remove the gearchange pedal and the crankcase left-hand outer cover, then roll off the spring retainer to free the rubber gaiter from the end of the driveshaft casing.

5 Remove the suspension unit lower mounting nuts. Knock back the locking tab which secures the pivot shaft head, then unscrew and withdraw the pivot shaft, supporting the swinging arm as it comes free. The swinging arm assembly can now be pulled rearwards until the universal joint splines separate and lifted away.

6 Examine the pivot shaft for wear or damage, renewing it if obviously worn or bent. The bushes in the frame cross-tube should be renewed if they are worn internally or if they have worked loose in the frame. In theory, the bushes can be drifted out and the new ones tapped into position. In practice, they can be notoriously difficult to remove, especially if they have corroded in place.

7 Try driving the old bushes out, using a socket as a drift. If this fails, make up a drawbolt arrangement to pull the old bushes out, or as a last resort, hacksaw through the bush, taking care not to cut into the frame. The bush can then be collapsed and driven out. When fitting new bushes, grease the outer surface, and make sure that they enter the frame squarely. Note that in rare instances where play develops between the pivot shaft and the swinging arm it will be necessary to renew the swinging arm casting, unless a local engineering works can undertake welding and re-machining of the worn bore and thread.

8 When refitting the swinging arm, tighten the pivot nut to 5.6 kgf m (40 lbf ft). Do not omit to bend up the locking tab to retain the pivot shaft head. When installing the rear wheel, tighten the rear wheel spindle nut to 6.0 kgf m (43 lbf ft). Fit a new split pin to retain the wheel spindle nut. Refit the brake torque arm, tightening the nut to 1.5 kgf m (11 lbf ft). Refit the suspension units, tightening the upper mounting bolts to 3.0 kgf m (22 lbf ft) and the lower mounting bolts to 2.3 kgf m (17 lbf ft). Check and adjust the rear brake pedal free play and rear brake switch adjustment.

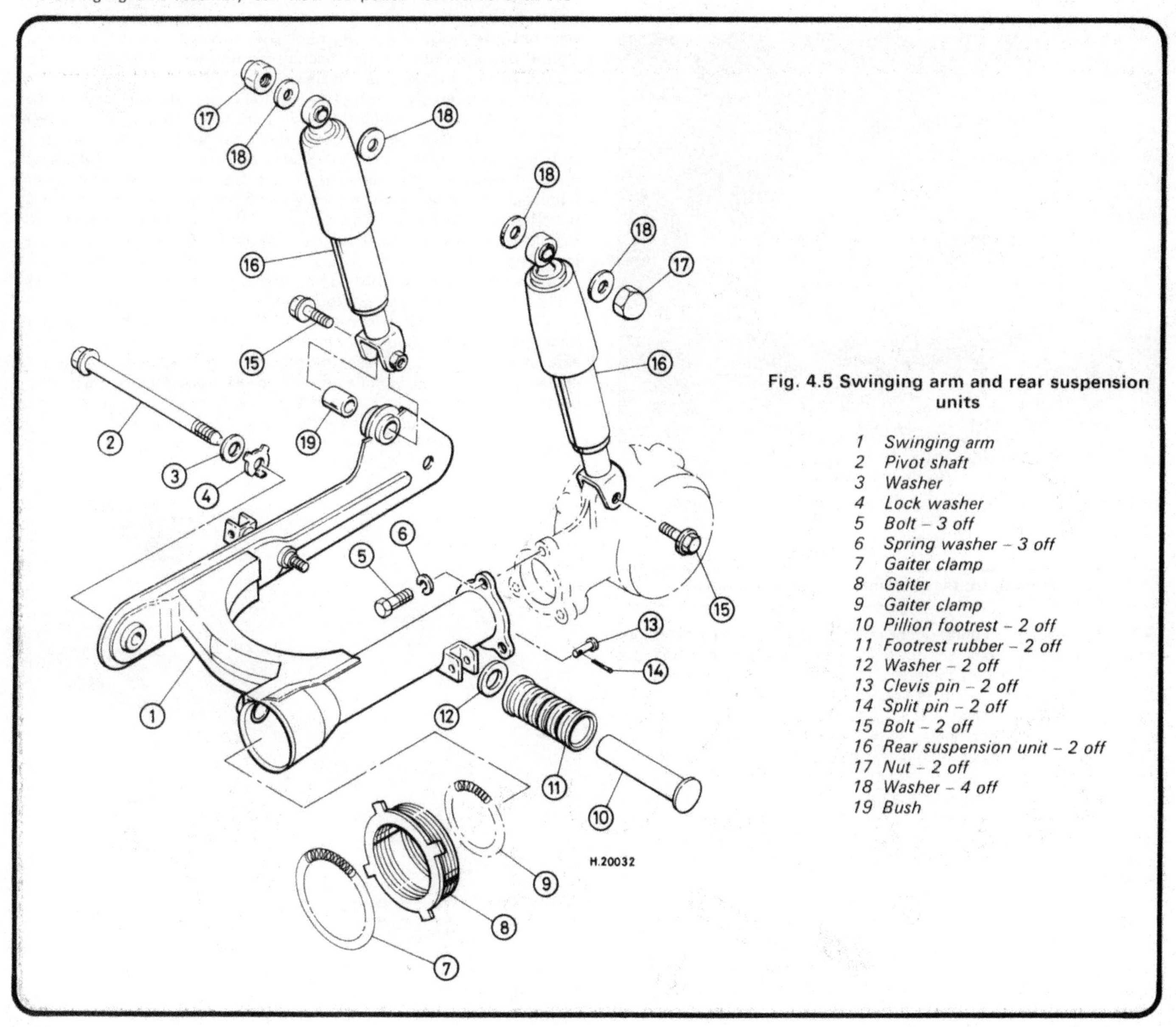

Fig. 4.5 Swinging arm and rear suspension units

1 Swinging arm
2 Pivot shaft
3 Washer
4 Lock washer
5 Bolt – 3 off
6 Spring washer – 3 off
7 Gaiter clamp
8 Gaiter
9 Gaiter clamp
10 Pillion footrest – 2 off
11 Footrest rubber – 2 off
12 Washer – 2 off
13 Clevis pin – 2 off
14 Split pin – 2 off
15 Bolt – 2 off
16 Rear suspension unit – 2 off
17 Nut – 2 off
18 Washer – 4 off
19 Bush

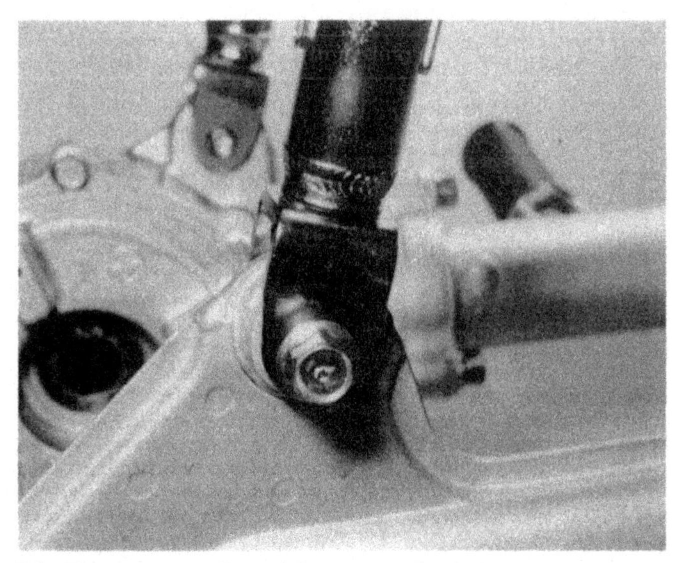

8.5a Remove suspension unit lower mounting bolts

8.5b Flatten locking tab and unscrew the pivot shaft

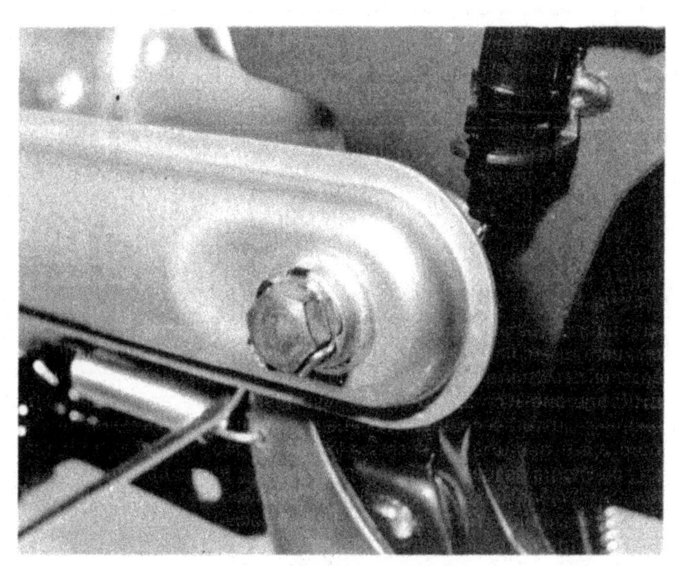

8.8 Tighten pivot shaft, securing the head with the locking tab

9 Rear suspension units: examination and renovation

1 The rear suspension units can be removed by unscrewing the upper and lower mounting bolts. For obvious reasons, deal with one unit at a time or the back of the machine will collapse onto the rear wheel.
2 The springs can be removed from the units, but only if the correct spring compressor is available. It is suggested that this job is entrusted to an authorized Yamaha dealer, since the threaded ends of the units will be damaged if tackled incorrectly. If facilities permit dismantling, note that the service limit for the spring free length is 168 mm (6.6 in).
3 Check the units for signs of leakage. If either unit leaks or does not operate smoothly, renew them as a pair. When refitting the units, note that the upper mounting bolts should be tightened to 3.0 kgf m (22 lbf ft) and the lower mounting bolts to 2.3 kgf m (17 lbf ft).

10 Shaft drive assembly: dismantling and reassembly

1 The shaft drive system employed on the Yamaha T80 models is intended to eliminate the normal maintenance requirements of conventional chain drive. In normal use the shaft assembly and bevel gears will give many years service without attention, but if the drive train becomes noisy or excessive free play develops, the swinging arm should be removed from the machine so that the driveshaft and the various gears and bearings can be checked.
2 With the assembly removed and on the bench, slacken and remove the three bolts which retain the shaft housing to the rear wheel bevel gearbox. The bevel gearbox can now be pulled away from the swinging arm, together with the driveshaft. This in turn can be pulled out of the end of the bevel gearbox. Moving back to the engine unit, remove the three bolts which secure the middle gear assembly bearing plate to the middle gear case extension of the crankcase. Lift away the middle gear shaft, together with the bearing plate and the universal joint. The drive train is now separated in to three main areas, the middle gear assembly and universal joint, the driveshaft and transmission shock absorber, and the bevel gearbox.
3 Check the driveshaft components as described in the following Sections. Refit the middle gear assembly into the middle gearbox extension on the crankcase, having first checked the gear tooth clearance. Fit the appropriate shim(s) to the bearing plate, then offer up the assembly, making sure that the teeth mesh correctly. Fit the three retaining bolts and tighten them to 1.0 kgf m (7.2 lbf ft).
4 Fit the coupling sleeve over the projecting end of the bevel gearbox drive pinion, grease the inside of the coupling, then engage the

10.2a Remove bolts and separate the bevel gearbox from the swinging arm unit

10.2b The shaft assembly can be withdrawn from the bevel gearbox

10.2c Middle gear assembly is retained to crankcase by three bolts

driveshaft. Offer up the bevel gearbox and shaft assembly to the swinging arm unit. Fit the retaining bolts, tightening them to 2.6 kgf m (19 lbf ft). Continue reassembly by refitting the swinging arm assembly into the frame. Refit the rear wheel, remembering to check rear brake adjustment and the brake lamp switch setting.
5 In the event of leakage from the bevel gearbox, check that it has not been over-filled with grease, and that the cover O-ring is intact. If the problem persists, have the assembly checked by a Yamaha dealer, who may be able to offer a modification to resolve the problem.

11 Middle gear assembly: examination and renovation

1 Start by examining the assembly for signs of wear or damage. Feel for free play in the universal joint bearings. If movement can be felt, or if the bearings feel stiff or gritty when moved, dismantle the joint so that they can be checked in more detail. It is preferable not to dismantle the joint unless the bearing condition is suspect.
2 Start by removing the circlips which retain the bearings in the joint bores. There are a total of four bearings and circlips, but it is best to deal with each opposing pair of bearings separately. Find a tubular spacer with an inside diameter slightly larger than that of the bearing (an inverted socket will be quite adequate for this). Place the spacer on the workbench, then position the joint so that one of the bearings is immediately above it.
3 Using a drift of suitable diameter (a socket will do for this too) tap the uppermost bearing down, displacing the joint and thus the lower bearing down and into the spacer. Once the bearing has partially emerged from its bore, grasp it with a pipe wrench or similar and pull it clear. As it comes free, the bearing needle rollers may drop free; take care to retain all of them. Turn the assembly over and tap it with a hammer to dislodge the second bearing, then separate the joint halves. The process should now be repeated on the remaining pair of bearings, and the yoke removed to inspect its bearing surfaces.
4 Clean the bearing outer races, the bearing needle rollers and the yoke. Examine the assembly for wear, scoring or cracked or flattened rollers. If wear or damage is found, or if there was detectable free play in any bearing when it is fitted over the bearing surface of the yoke, Yamaha recommend that the bearings, yoke and the driveshaft half of the joint be renewed. If the assembly is in good condition it can be reassembled and re-used. Before reassembly commences, note that if the shaft bearing, shaft or bearing plate require attention, now is the time to do it; the universal joint must be separated to permit access to these items.
5 Clamp the universal joint end of the middle gear shaft between soft jaws in a vice. Slacken and remove the large retaining nut (it will be tight) and displace and remove the shaft from the bearing. Check the bearing for wear or damage. If it seems sloppy, or feels stiff or gritty when it is turned, it should be renewed. The bearing is secured in the bearing plate by a threaded retainer with an internal hexagon. To remove this a special 27 mm hexagon wrench is needed (Yamaha Part Number 90890-01363). Alternatively, an old truck wheel nut can be filed down to fit (see photograph). Clamp the bearing plate between soft jaws, then unscrew and remove the retainer. The bearing can be tapped out of the retainer in the normal way, noting that warming the assembly in nearly boiling water will make this operation much easier.
6 When refitting the bearing retainer, use Loctite 242 or similar on the threads, and tighten it to 6 kgf m (43 lbf ft). Refit the shaft and the universal joint half, apply Loctite to the retaining nut and tighten to 9 kgf m (65 lbf ft).
7 The universal joint must be assembled very carefully. Pack the bearing outer races with grease, using it to hold the needle rollers in place. As each bearing is fitted over the yoke, make sure that none of the rollers are displaced. If this happens, the roller will tend to get pushed to the bottom of the outer race, so if the bearing will not seat fully over the yoke, dismantle and check this. The bearing outer races can be pressed into the joint bores using a socket of suitable diameter and a vice. Once in place, fit the circlips, ensuring that they locate fully.
8 The middle gear assembly must be checked to ensure that the gear tooth mesh depth is set accurately. This is done by measuring the length 'L' in the accompanying line drawing. Place a straight-edge across the end face of the driven gear, then use a vernier caliper to measure between it and the bearing plate gasket face. The required thickness of shims to be fitted to the bearing plate is then calculated according to the measured length, using the table below.

Length 'L'	Shim thickness required
37.83 – 37.90 mm	0.4 mm
37.90 – 38.00 mm	0.5 mm
38.00 – 38.10 mm	0.6 mm
38.10 – 38.20 mm	0.7 mm
38.20 – 38.30 mm	0.8 mm
38.30 – 38.40 mm	0.9 mm
38.40 – 38.45 mm	1.0 mm

9 Having worked out the required thickness of shims, select the smallest number of shims from the following sizes to make up this thickness. Shims are available in 0.15 mm, 0.20 mm, 0.30 mm, 0.40 mm and 0.50 mm thicknesses. Fit the appropriate shim(s) to the bearing plate, then refit the assembly, making sure that the teeth mesh correctly. Fit the three retaining bolts and tighten them to 1.0 kgf m (7.2 lbf ft).

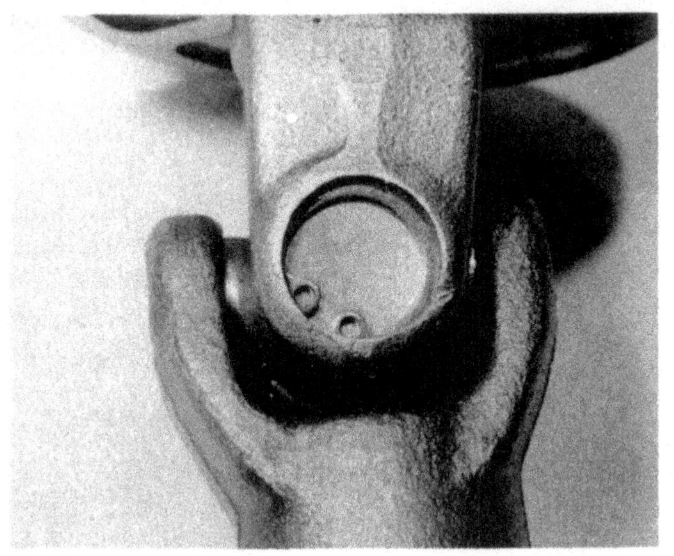
11.2 Universal joint bearings are secured by circlips

11.3 Dismantled universal joint, showing needle roller bearings

11.5a Slacken and remove the universal joint retaining nut ...

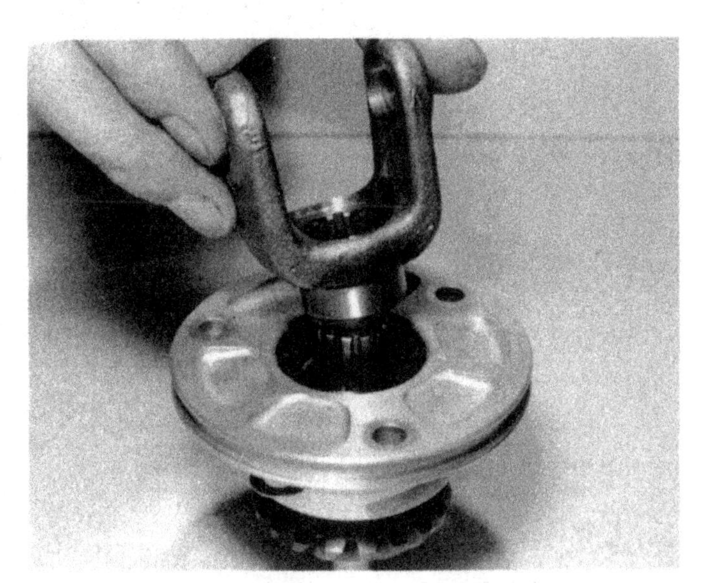
11.5b ... and lift the joint half away from the shaft end

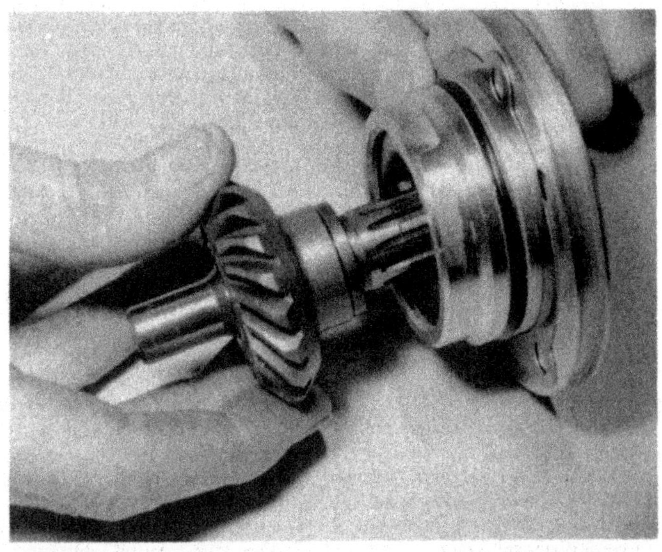
11.5c The shaft can now be displaced from the bearing assembly

11.5d Unscrew the bearing retainer, then drive bearing out if renewal is required

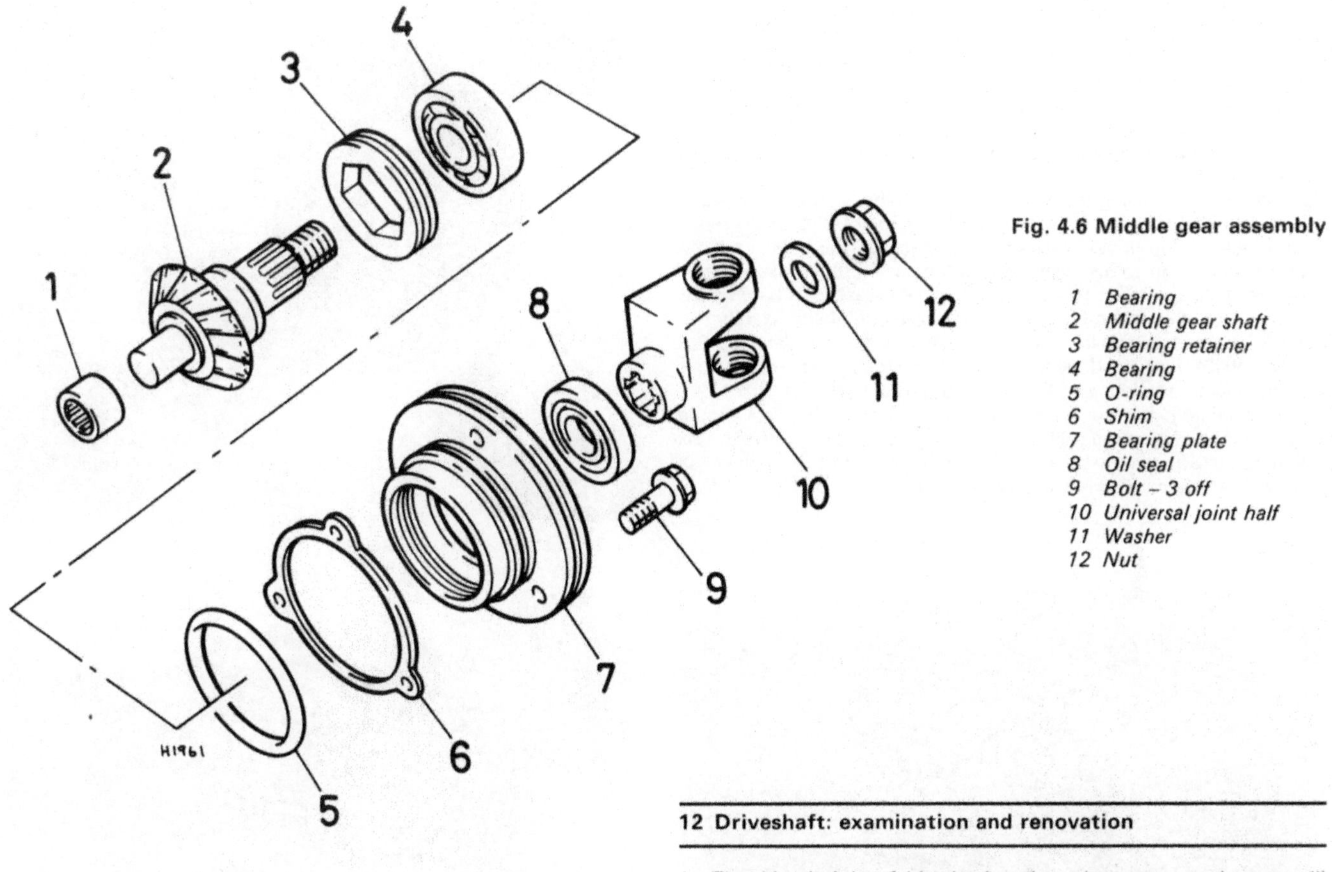

Fig. 4.6 Middle gear assembly

1 Bearing
2 Middle gear shaft
3 Bearing retainer
4 Bearing
5 O-ring
6 Shim
7 Bearing plate
8 Oil seal
9 Bolt – 3 off
10 Universal joint half
11 Washer
12 Nut

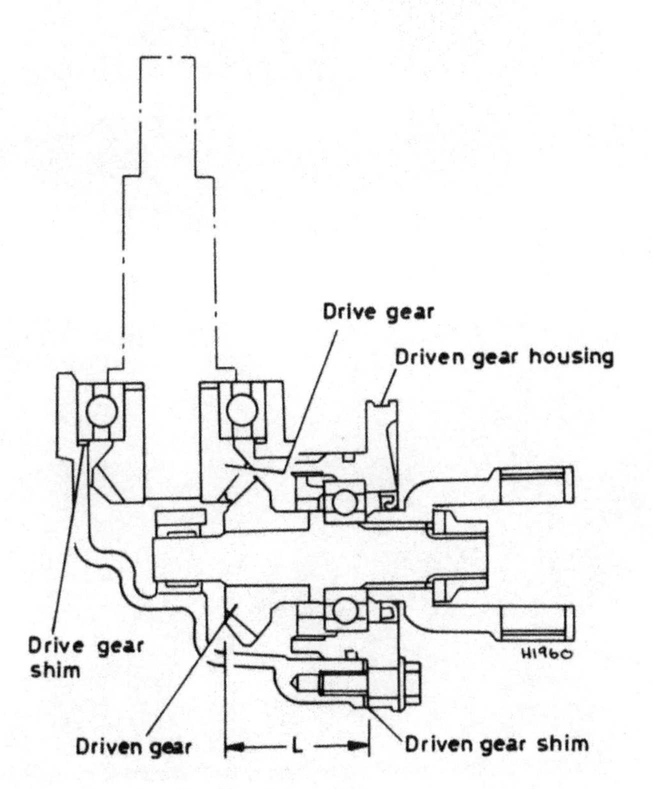

Fig. 4.7 Middle gear assembly gear tooth mesh depth measurement

12 Driveshaft: examination and renovation

1 The driveshaft is a fairly simple unit, and any wear or damage will be quite evident during examination. The spring and helical spline arrangement at the driving end form a transmission shock absorber which reduces the level of transmission snatch when moving off or under braking. If the shock absorber components require attention, a safe method of compressing the spring must be found, otherwise this stage should be left to an authorized Yamaha dealer.
2 With the spring compressed, remove the circlip, followed by the spacer, helically-splined boss, washer, spring, spring retainer and the second circlip. If wear is found, renew the affected parts. Wear is most likely to affect the internal and external splines of the boss, the corresponding splines of the driveshaft and the spring. If wear is widespread it may be preferable to renew the shaft assembly complete. The shaft shock absorber components are refitted by reversing the dismantling sequence, using grease on the boss splines. Before the shaft is installed, grease the splines at each end.

13 Bevel gearbox: examination and renovation

1 The bevel gearbox can be dealt with after the swinging arm assembly has been removed as described in Section 8. Alternatively, remove the rear wheel and unbolt the gearbox from the back of the swinging arm casing.
2 Remove the circlip from the centre of the cush drive hub and lift the hub away. Slacken and remove the six bolts which secure the bearing plate, lifting it away to reveal the large driven gear and the smaller drive gear.
3 At this stage it is possible to assess the general condition of the gear teeth. If the gears are to be removed, or if access is required to the drive gear bearing or the driven gear left-hand bearing, proceed as follows.
4 Using the 32 mm hexagon wrench (Yamaha Part Number 90890-01364) or alternatively a larger version of the filed-down wheelnut described in Section 11, slacken and remove the bearing retainer. Displace and remove the drive gear assembly and its bearing. If it proves tight, warm the casing using near boiling water to expand the alloy.

5 Examine the drive and driven gear teeth for wear or damage, renewing them as required. Check the drive gear front bearing for wear and renew it if is obviously worn or if it is noisy when turned. If the drive gear rear bearing is worn, the complete gear assembly must be renewed; the bearing is not supplied separately. The driven gear bearings can be checked for wear in the normal way. If renewal is necessary, drive the old bearing out using a socket as a drift. Note that the area around the bearing boss should be well supported during removal and fitting.

6 Prior to reassembling the gearbox, clean off the old grease from the housing and bevel gears. Apply 120 g of fresh Shell TG093 #00 grease to the teeth of the driven gear and fit the gear in the housing. Note that it is important to avoid over-filling the casing otherwise leakage may result. As the exact type and quantity of grease is of paramount importance in this assembly, a tube of grease containing exactly the correct quantity is available from Yamaha dealers under Part No 90793 - 40001. Refit the drive gear, followed by the spacer and the front bearing. Fit the retainer and tighten it to 6.0 kgf m (43 lbf ft).

7 Refit the large bearing plate, using a new O-ring. The plate should be positioned so that the cast-in arrow mark faces forward. Fit the six retaining bolts and tighten them to 1.0 kgf m (7.2 lbf ft). Refit the cush drive hub and secure it with its circlip.

13.2a Cush drive hub is retained by a circlip (arrowed)

13.2b Remove hub, then release cover bolts ...

13.2c ... and remove cover to reveal the bevel gearbox components

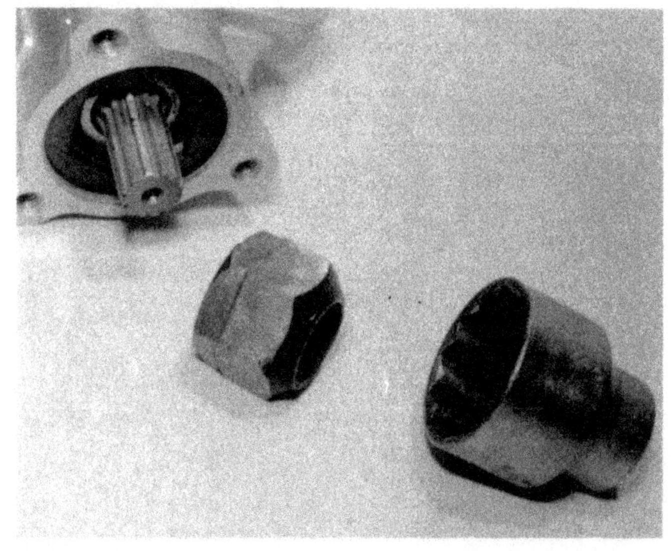

13.4a Improvised retainer wrench made from old truck wheelnut

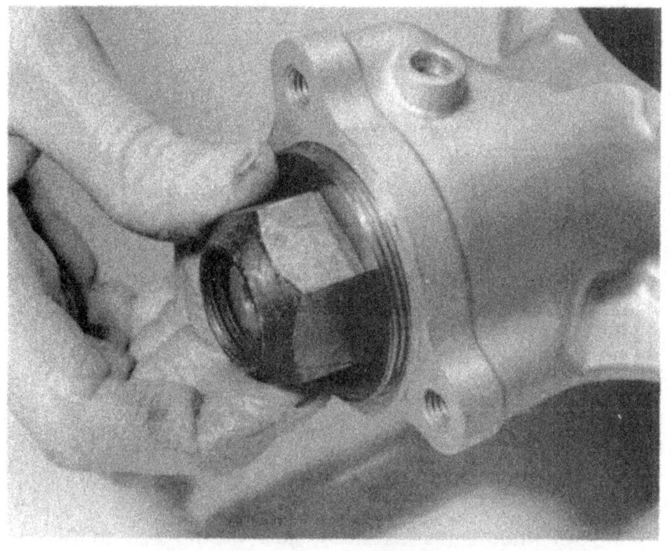

13.4b Place tool over shaft end to engage internal hexagon ...

13.4c ... clamp housing between wooden blocks in a vice and slacken the retainer

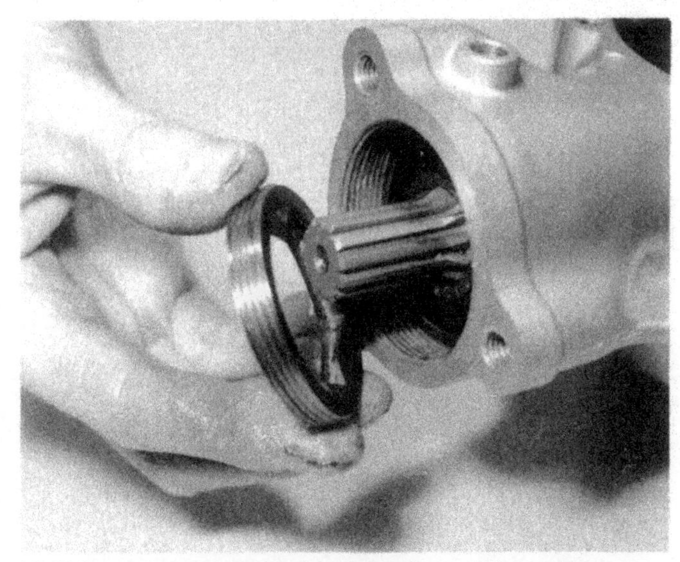

13.4d Remove the retainer to free the shaft assembly and bearings

13.5a Drive gear front bearing is easily removed and renewed – rear bearing is supplied as an assembly with shaft

13.5b Driven gear inner bearing requires an extractor to remove it from its blind bore

13.6a Check condition of O-ring on driven gear ...

13.6b ... then install in casing and apply the specified grease to the teeth

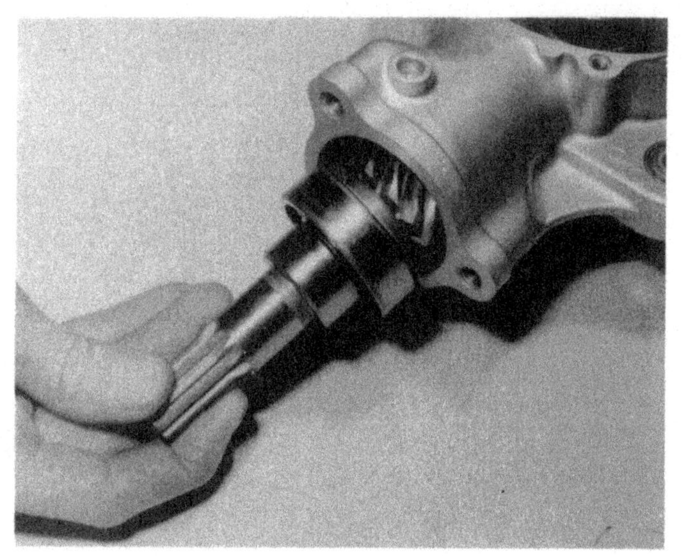

13.6c Fit the drive gear into the casing ...

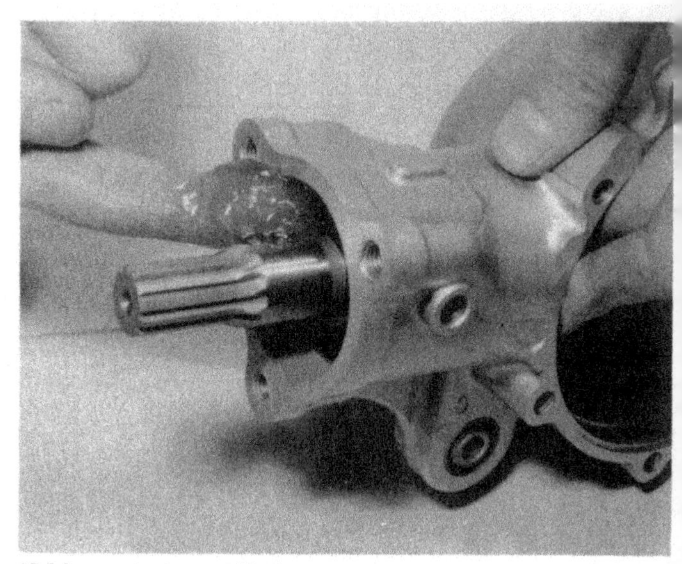

13.6d ... apply the specified grease ...

13.6e ... and fit the bearing spacer

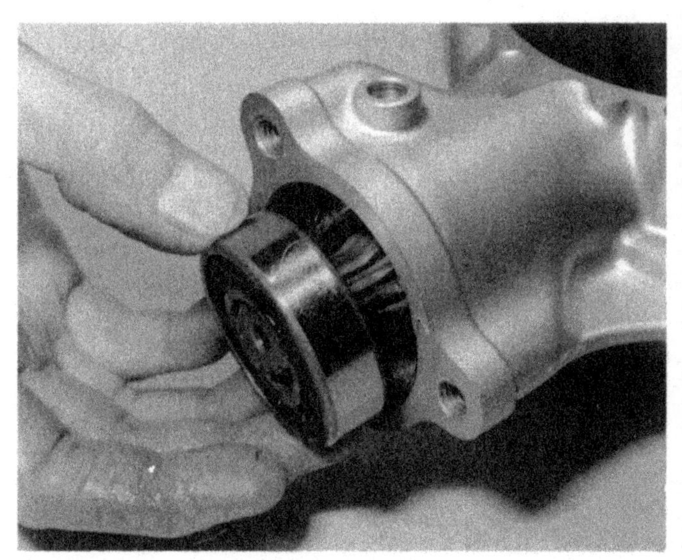

13.6f Install the bearing, then refit the retainer

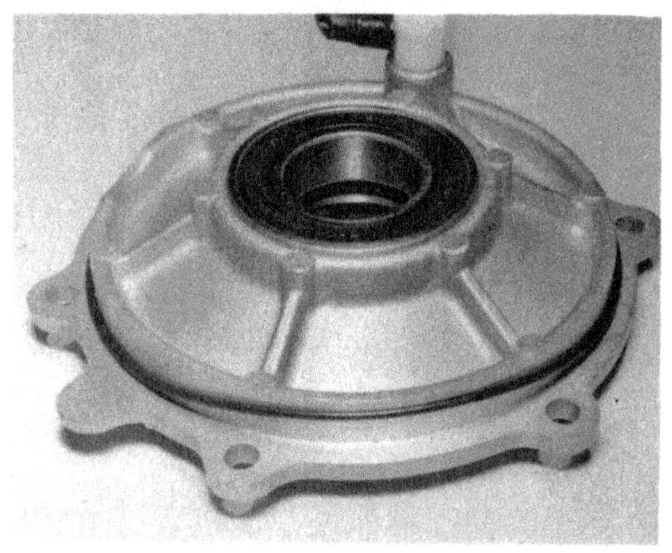

13.7a Fit housing cover, using a new O-ring

13.7b Tighten bolts evenly. Arrow mark must face front

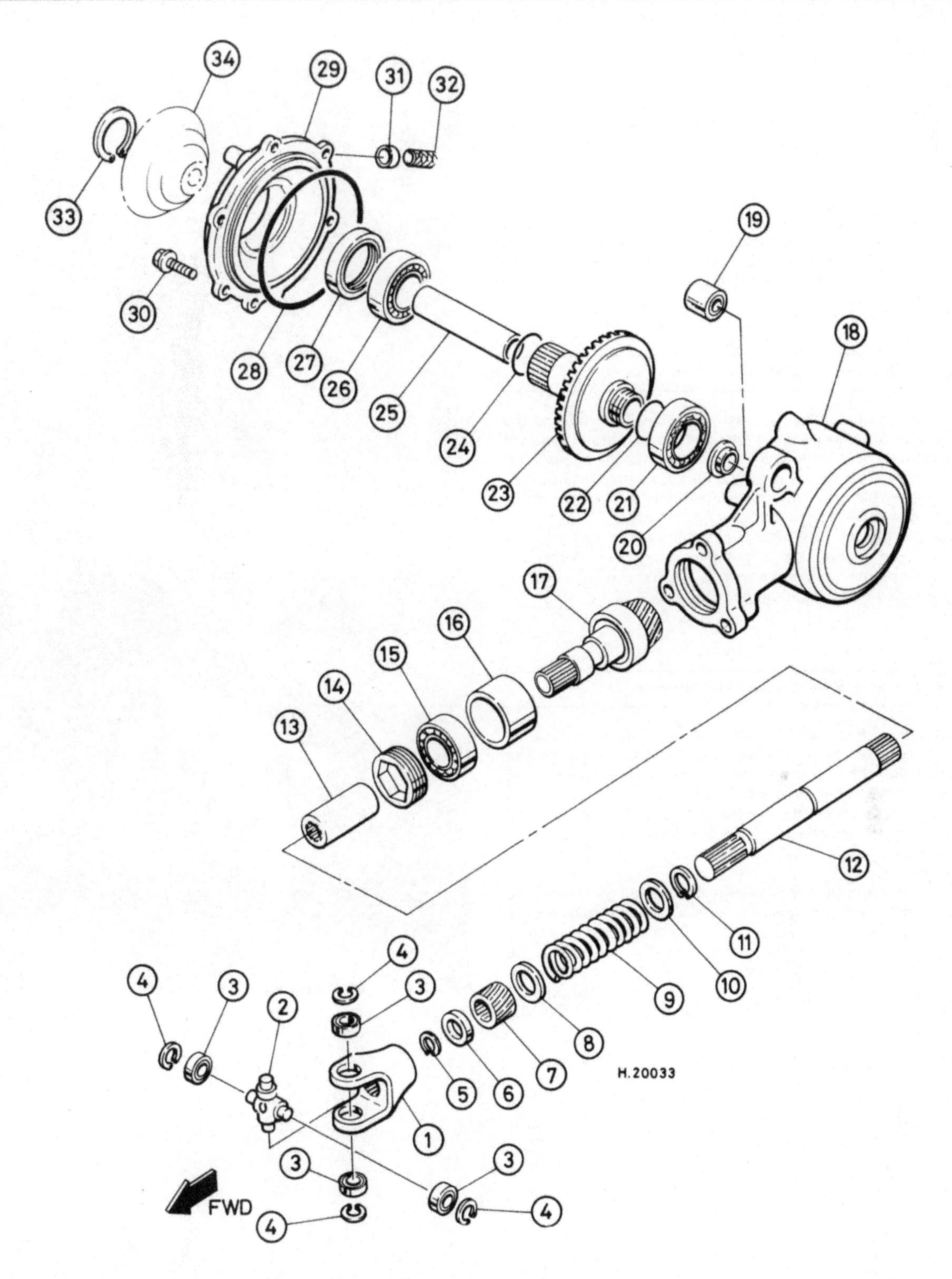

Fig. 4.8 Driveshaft and bevel box

1 *Universal joint half*
2 *Yoke*
3 *Bearing – 4 off*
4 *Circlip – 4 off*
5 *Circlip*
6 *Spacer*
7 *Helically-splined boss*
8 *Washer*
9 *Spring*
10 *Spring retainer*
11 *Circlip*
12 *Driveshaft*
13 *Coupling sleeve*
14 *Bearing retainer*
15 *Driveshaft front bearing*
16 *Bearing spacer*
17 *Drive gear and rear bearing*
18 *Casing*
19 *Suspension bush*
20 *Spacer*
21 *Driven gear inner bearing*
22 *O-ring*
23 *Driven gear*
24 *O-ring*
25 *Spacer*
26 *Driven gear outer bearing*
27 *Oil seal*
28 *O-ring*
29 *Bearing plate*
30 *Bolt – 6 off*
31 *Clip**
32 *Hose**
33 *Circlip*
34 *Cush drive hub*

**Fitted to 2FL model only*

14 Stands: examination and renovation

1 The centre stand is attached to the bottom of the frame main spine, retained by a headed spindle which is itself secured by a circlip. A tension spring holds the stand retracted when not in use. The side stand (where fitted) is held by a shouldered pivot bolt and locknut to the footrest bar.
2 Both stands should be checked regularly for security, and the pivots lubricated. Check that the return springs are sound, and renew them if there are signs of cracking or serious rusting. In the case of the side stand, do not fit an ordinary bolt in place of the correct hardened and shouldered type or it may shear off in use.

14.2 Check condition of stands and springs. Lubricate the pivots

15 Footrests: examination and renovation

1 The front footrest assembly is attached to the underside of the crankcase by three bolts and is easily removed for painting or repair. If the footrests become bent, remove them from the machine and pull off the footrest rubbers. Heat up the damaged area with a blowlamp, then hammer or bend the bar back to its normal position. It follows that the assembly will need to be repainted following the repair.
2 The passenger footrests are individual folding items held in place by a clevis pin and secured by a split pin. Apart from occasional lubrication of the pivots, no maintenance is required. In the event of damage, the affected footrest should be renewed.

16 Speedometer and drive cable: maintenance

1 The speedometer head is a sealed unit mounted in the upper section of the handlebar nacelle. It is held in position by a wire clip and can be removed after the headlamp assembly has been detached (two screws) and the drive cable removed. The clip is rather awkward to release with the top section of the headlamp nacelle in position, but it can be done with a little patience. The accompanying photograph shows the nacelle dismantled so that the position of the clip can be seen more easily.
2 The speedometer is driven by a flexible cable from a gearbox in the front wheel. In the event of failure, it is almost certainly caused by a broken cable, so always check for this first. Jerky or erratic operation is often caused by a kink in the inner cable, and this can usually be remedied by fitting a new one. The inner cable can be withdrawn for lubrication or renewal. When lubricating it, do not grease the upper six inches or so to avoid excess grease finding its way into the speedometer head.

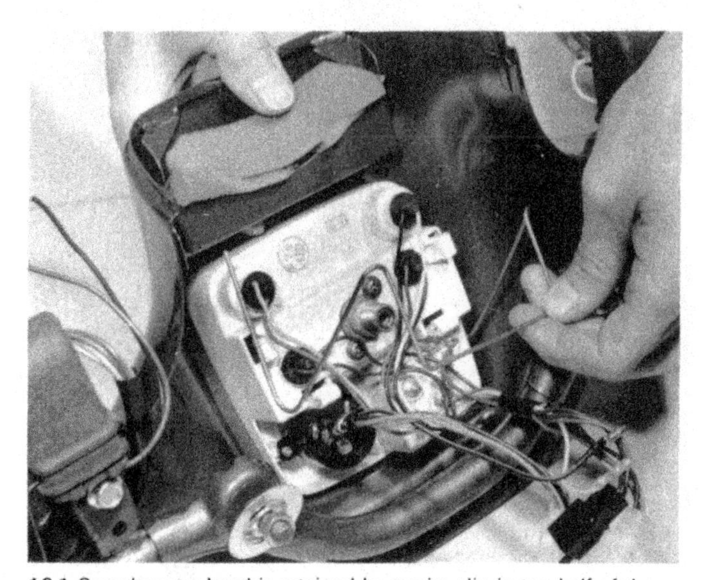
16.1 Speedometer head is retained by a wire clip in top half of the headlamp nacelle

Chapter 5 Wheels, brakes and tyres

Contents

Specifications

Wheels

Type	Wire spoked, chromium plated steel rims
Size	17 in, front and rear

Tyres

Size – front and rear:		
Early (35T) T80 model	2.50-17-4PR	
All T50 and later (2FL) T80 models	2.50-17-4PR (front), 2.50-17-6PR (rear)	
Tyre pressures (cold):	**Front**	**Rear**
Up to 75 kg (165 lb) load	22 psi (1.54 kg/cm²)	28 psi (2.0 kg/cm²)
Above 75 kg (165 lb) load – early (35T) T80 models	22 psi (1.54 kg/cm²)	32 psi (2.25 kg/cm²)
Above 75 kg (165 lb) load – all T50 and later (2FL) T80 models	22 psi (1.54 kg/cm²)	40 psi (2.8 kg/cm²)

Brakes

Type:	
Front	Single leading shoe drum brake, operated from handlebar lever
Rear	Single leading shoe drum brake, operated from foot pedal
Drum diameter (front and rear)	110 mm (4.33 in)
Service limit	111 mm (4.37 in)
Lining thickness (front and rear)	4.0 mm (0.157 in)
Service limit	2.0 mm (0.08 in)
Return spring free length (front and rear)	54 mm (2.13 in)
Front brake lever free play	5.0 – 8.0 mm (0.2 – 0.3 in)
Rear brake pedal free play	20 – 30 mm (0.8 – 1.2 in)

Torque wrench settings

Component	kgf m	lbf ft
Front wheel spindle	3.9	28.0
Rear wheel spindle	6.0	43.0
Rear brake torque arm	1.5	11.0
Brake arm pinch bolt	0.7	5.1

1 General description

The Yamaha T50 and 80 models are equipped with wire-spoked wheels having chromium plated steel rims, front and rear. Each wheel carries a tubed tyre. Braking is by single leading shoe (sls) drum on both wheels, the front brake being operated by a lever on the right-hand side of the handlebar, whilst the rear brake is controlled by a foot pedal.

2 Wheels: examination and renovation

1 Wire spoked wheels are often viewed as being prone to problems when compared to the increasingly popular cast alloy and composite tyres. Whilst this is true to some extent, it is also true that wire spoked wheels are relatively easy and inexpensive to adjust or repair. Spoked wheels can go out of true over periods of prolonged use and like any

wheel, as the result of an impact. The condition of the hub, spokes and rim should therefore be checked at regular intervals.
2 For ease of use an improvised wheel stand is invaluable, but failing this the wheel can be checked whilst in place on the machine after it has been raised clear of the ground. Make the machine as stable as possible, if necessary using blocks beneath the crankcase as extra support. Spin the wheel and ensure that there is no brake drag. If necessary, slacken the brake adjuster until the wheel turns freely.
3 Slowly rotate the wheel and examine the rim for signs of serious corrosion or impact damage. Slight deformities, as might be caused by running the wheel along a curb, can often be corrected by adjusting spoke tension. More serious damage may require a new rim to be fitted, and this is best left to an expert. Whilst this is not an impossible undertaking at home, there is an art to wheel building, and a professional wheel builder will have the facilities and parts required to carry out the work quickly and economically. Badly rusted steel rims should be renewed in the interests of safety as well as appearance. Where light alloy rims are fitted corrosion is less likely to be a serious problem, though neglect can lead to quite substantial pitting of the alloy.
4 If it has been decided that a new rim is required some thought should be given to the size and type of the replacement rim. In some instances the problem of obtaining replacement tyres for an oddly sized original rim can be resolved by having a more common rim size fitted. Do check that this will not lead to other problems, fitting a new rim whose tyre fouls some other part of the machine could prove a costly error. Remember that changing the size of the rear wheel rim will alter the overall gearing. In most cases it should be possible to have a light alloy rim fitted in place of an original plated steel item. This will have a marginal effect in terms of weight reduction, but will prove far more corrosion resistant.
5 Assuming the wheel to be undamaged it will be necessary to check it for runout. This is best done by arranging a temporary wire pointer so that it runs close to the rim. The wheel can now be turned and any distortion noted. Check for lateral distortion and for radial distortion, noting that the latter is less likely to be encountered if the wheel was set up correctly from new and has not been subject to impact damage.
6 The rim should be no more than 2.0 mm (0.1 in) out of true in either plane. If a significant amount of distortion is encountered check that the spokes are of approximately equal tension. Adjustment is effected by turning the square-headed spoke nipples with the appropriate spoke key. This tool is obtainable from most good motorcycle shops or tool retailers.
7 With the spokes evenly tensioned, any remaining distortion can be pulled out by tightening the spokes on one side of the hub and slackening the corresponding spokes from the opposite hub flange. This will allow the rim to be pulled across whilst maintaining spoke tension.
8 If more than slight adjustment is required it should be noted that the tyre and inner tube should be removed first to give access to the spoke ends. Those which protrude through the nipple after adjustment should be filed flat to avoid the risk of puncturing the tube. It is essential that the rim band is in good condition as an added precaution against chafing. In an emergency, use a strip of duct tape as an alternative; unprotected tubes will soon chafe on the nipples.
9 Should a spoke break a replacement item can be fitted and retensioned in the normal way. Wheel removal is usually necessary for this operation, although complete removal of the tyre can be avoided if care is taken. A broken spoke should be attended to promptly because the load normally taken by that spoke is transferred to adjacent spokes which may fail in turn.
10 Remember to check wheel condition regularly. Normal maintenance is confined to keeping the spokes correctly tensioned and will avoid the costly and complicated wheel rebuilds that will inevitably result from neglect. When cleaning the machine do not neglect the wheels. If the rims are kept clean and well polished many of the corrosion related maladies will be prevented.

3 Front wheel: removal and refitting

1 Place the machine on its centre stand on level ground, then raise the front wheel clear of the ground by placing wooden blocks or a jack below the crankcase. Prise out the wire clip which secures the speedometer cable to the brake backplate. Slacken and remove the front brake cable adjuster nut and disengage the cable from the operating arm. Refit the trunnion and nut on the cable for safe keeping.
2 Straighten and remove the split pin from the wheel spindle nut. Slacken and remove the nut, then support the weight of the wheel while the wheel spindle is displaced and removed. The wheel can now be manoeuvred clear of the forks and placed to one side.
3 Reassembly is basically a reversal of the above sequence, noting the following points. Check that the front brake backplate is fitted correctly and that the speedometer drive engages correctly over the hub slots. Offer up the front wheel, aligning the slot in the brake backplate with the boss on the left-hand suspension link. Fit the wheel spindle (from the right-hand side), then fit and tighten the wheel spindle nut to 3.9 kgf m (28 lbf ft). Reconnect the speedometer cable and the front brake cable. Set the brake cable adjuster to give 5 – 8 mm (0.2 – 0.3 in) free play, measured at the lever end.

3.1a Speedometer cable is secured by a wire clip at the wheel end

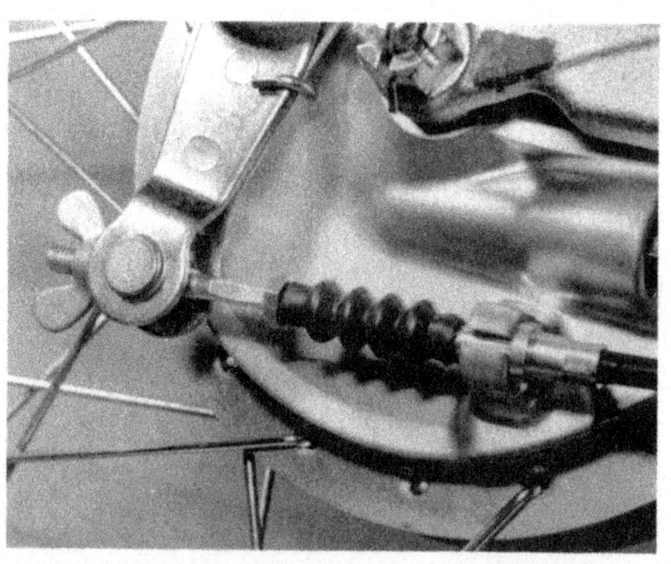

3.1b Unscrew the brake cable adjuster to free the cable from the brake arm

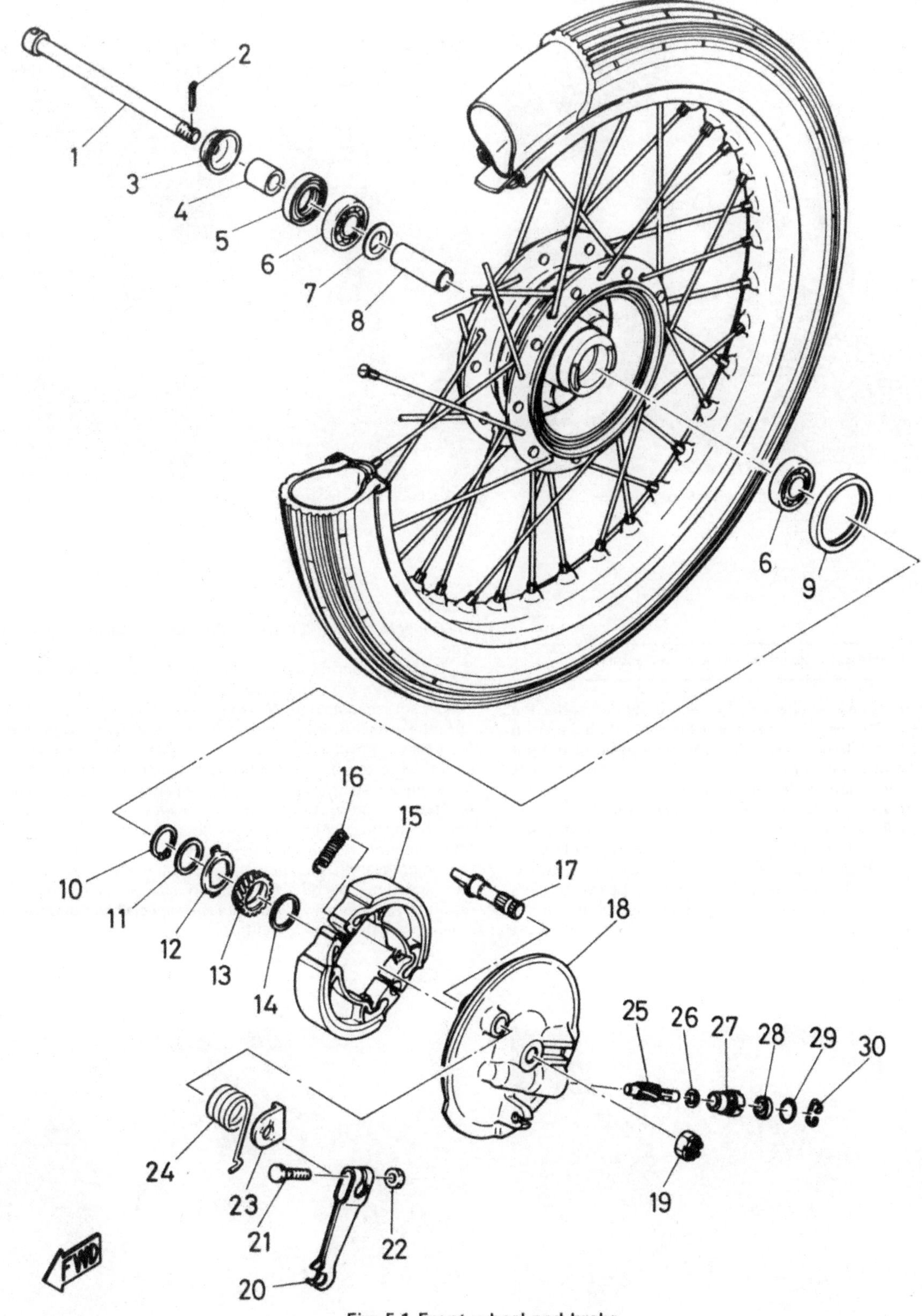

Fig. 5.1 Front wheel and brake

1	*Spindle*	*9*	*Grease seal*	*16*	*Return spring – 2 off*	*24*	*Return spring*
2	*Split pin*	*10*	*Circlip*	*17*	*Brake cam*	*25*	*Speedometer cable driven gear*
3	*Dust seal*	*11*	*Washer*	*18*	*Brake backplate*	*26*	*Washer*
4	*Spacer*	*12*	*Speedometer drive plate*	*19*	*Nut*	*27*	*Bush*
5	*Grease seal*	*13*	*Speedometer cable drive gear*	*20*	*Brake arm*	*28*	*Oil seal*
6	*Bearing – 2 off*	*14*	*Washer*	*21*	*Pinch bolt*	*29*	*O-ring*
7	*Spacer flange*	*15*	*Brake shoe – 2 off*	*22*	*Nut*	*30*	*Clip*
8	*Spacer*			*23*	*Wear indicator plate*		

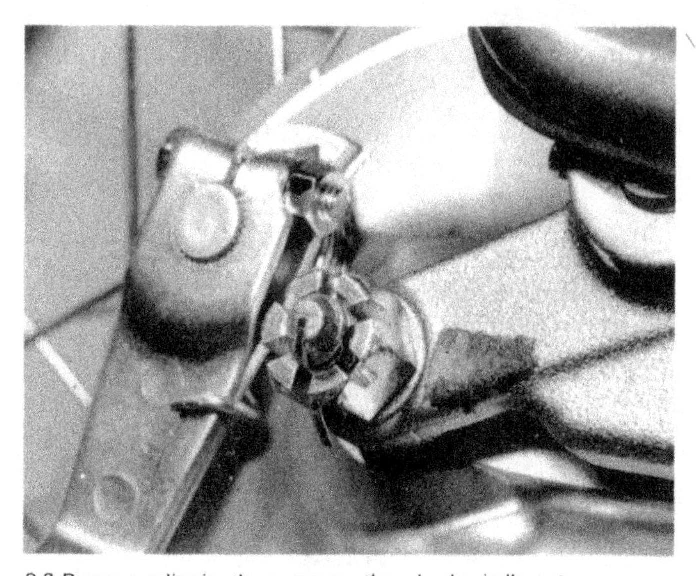

3.2 Remove split pin, then unscrew the wheel spindle nut

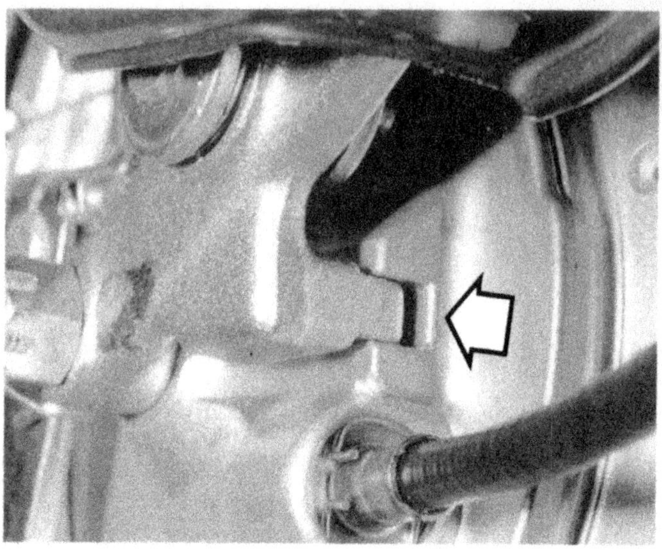

3.3 Check that the brake backplate engages over stop (arrowed)

4 Front whéel bearings: examination and renewal

1 The wheel bearings can be checked for wear or damage after the wheel has been removed from the machine as described above. Turn each bearing inner race with a finger, feeling for any roughness or tight spots. If the bearing does not turn smoothly and evenly, or if there is obvious free play in it, renèw both bearings as a set.
2 The bearings locate in bores in the light alloy hub unit and are separated by a tubular spacer. Start by removing the dust seal and spacer from the right-hand side of the hub, using a screwdriver to prise the seal out. It will probably be damaged during removal and should therefore be renewed during reassembly. To remove the bearings, place the wheel on its side on the workbench, supporting the hub on blocks so that the lower bearing can be driven out. Pass a long drift through the upper bearing and displace the tubular spacer by levering it to one side.
3 Lodge the end of the drift against the edge of the inner race of the lower bearing, then drift it downwards, working around the bearing to keep it square in its bore. When it comes free, remove it and the spacer. Turn the wheel over and repeat the process on the remaining bearing. Note that removal will probably cause damage to even a good bearing; so do not re-use bearings once they have been removed from the hub.
4 Lubricate the new bearings by packing them with grease. Tap the first bearing into position, using a socket as a drift. Note that it is vital that the bearing enters the bore squarely, and that its sealed face is fitted outermost. Turn the wheel over and fit the spacer, then install the remaining bearing. Fit a new dust seal on the left-hand side of the hub, and remember to refit the headed spacer which sits in the seal. Carefully remove any residual grease, especially around the brake drum area, before the wheel is refitted.

4.4a Do not omit to fit the spacer between the wheel bearings

4.4b Fit bearings with sealed face outwards

4.4c Fit grease seal on right-hand side of the wheel ...

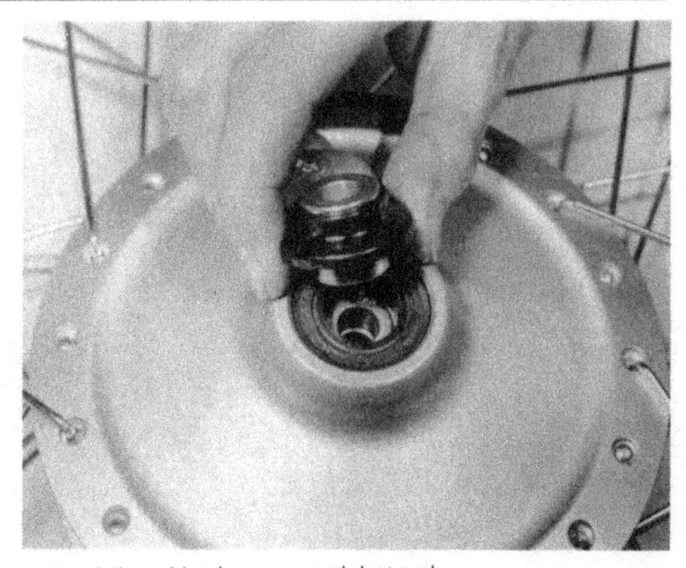
4.4d ... followed by the spacer and dust seal

5 Front brake: examination and renovation

1 The front brake backplate assembly can be lifted out of the drum for examination after the front wheel has been removed as described previously. Check the brake linings for scoring, contamination and wear. If the linings have become contaminated with oil or grease, the shoes should be renewed as a matter of course, but the source of the contamination should be identified and rectified before the new shoes are fitted. The lining material should not be less than 2.0 mm (0.08 in) thick at any point. If it has worn to, or beyond, this service limit, renew the shoes as a pair.
2 If the lining material is within limits, but has become glazed, it is permissible to roughen the surface with abrasive paper to restore braking efficiency. This job must be done outside, and a dust mask should be worn to avoid inhaling any of the dust, which contains asbestos. This applies equally when removing dust from the drum or the brake backplate.
3 The drum will not wear very quickly, but if it has reached the service limit of 111.0 mm (4.37 in) it must be renewed. If the drum surface has become scored it is permissible to have it skimmed on a lathe to correct the scoring, provided that this does not take the diameter beyond the above limit. Drum ovality is usually due to badly tensioned wheel spokes. Have this checked by an authorized Yamaha dealer or by a wheel specialist. As a last resort, this too can be corrected by skimming if it is not too severe. Fitting a new drum/hub assembly is a job for a professional wheel builder; it is not recommended that this be tackled at home.
4 To remove the shoes, turn the brake arm through 90° so that the shoes are pushed fully outwards by the cam. Remove the shoes by folding them inwards until they can be disengaged from the pivot and cam and the return springs released.
5 Remove the pinch bolt which retains the brake operating arm to the cam, lifting away the arm and its return spring. Lift off the wear indicator and the felt seal, then push the cam out of its bore in the backplate. Lift out the speedometer drive gear and the washer behind it, then prise out the dust seal.
6 Clean the brake backplate components thoroughly, removing all traces of old grease and brake dust. Allow the various parts to dry before reassembly commences. Before the new shoes are fitted, apply strips of masking tape or similar to the lining surface. This will prevent them from getting grease or oil spots on them during installation.
7 Grease the speedometer drive gear, and install it and its washer in the backplate recess. Fit the dust seal, using a new one if necessary. Wipe away any excess grease. Apply a trace of grease to the brake shoe pivot pin and to the operating cam. Refit the cam, followed by the felt seal. Fit the wear indicator, noting that this can be fitted in one position only. Fit the brake operating arm return spring followed by the arm, ensuring the dots on the cam and the arm coincide. Fit the pinch bolt and tighten it securely.
8 Assemble the new brake shoes and their return springs, then fit them by reversing the removal operation. Make sure that the shoes fit correctly over the pivot pin, and that they locate over the cam. Check that the drum is clean, then remove the protective tape from the linings before refitting the brake backplate into the wheel.

5.1 Brake backplate can be lifted away once wheel has been removed

5.2 Check linings for wear and contamination

5.7 Grease the speedometer drive before refitting backplate

6.1a Free the brake adjuster and rod from the brake arm

6 Rear wheel: removal and refitting

1 Remove the rear brake adjuster nut and the brake torque arm nut from the brake backplate. Disengage the brake operating rod from the arm, refitting the trunnion and nut on the end of the rod for safe keeping. Free the torque arm from the brake backplate.
2 Straighten and remove the split pin from the rear wheel spindle nut and slacken and remove the nut. Withdraw the wheel spindle. Displace the spacer on the right-hand side of the wheel, then pull the wheel away from the cush drive hub and manoeuvre it clear of the swinging arm.
3 The wheel can be refitted by reversing the removal sequence. Tighten the rear wheel spindle nut to 6.0 kgf m (43 lbf ft) and the brake torque arm nut to 1.6 kgf m (11 lbf ft). Secure the wheel spindle nut using a new split pin. Fit the R-pin which locks the torque arm nut, having renewed it if it is damaged. Unlike chain drive machines, there is no risk of the wheel being aligned incorrectly, and the need for adjustment of the wheel position does not apply. It is important to check, however, that the brake adjustment is set correctly. There should be 20 – 30 mm (0.8 – 1.2 in) free play at the pedal end. Check also that the rear brake lamp comes on just before the brake starts to take effect. If adjustment is needed, slacken the switch body nuts, and raise or lower the switch body until the lamp comes on at the required point. Tighten the locknuts.

6.1b Release brake torque arm

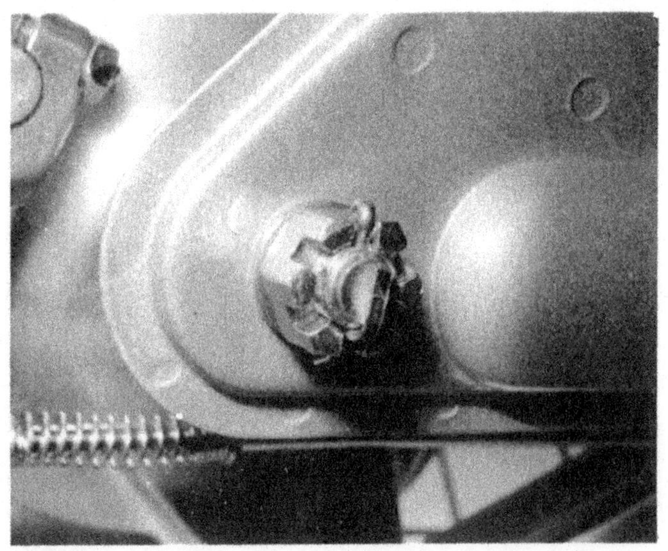

6.2a Remove split pin, then unscrew wheel spindle nut

6.2b Withdraw spindle and displace the spacer ...

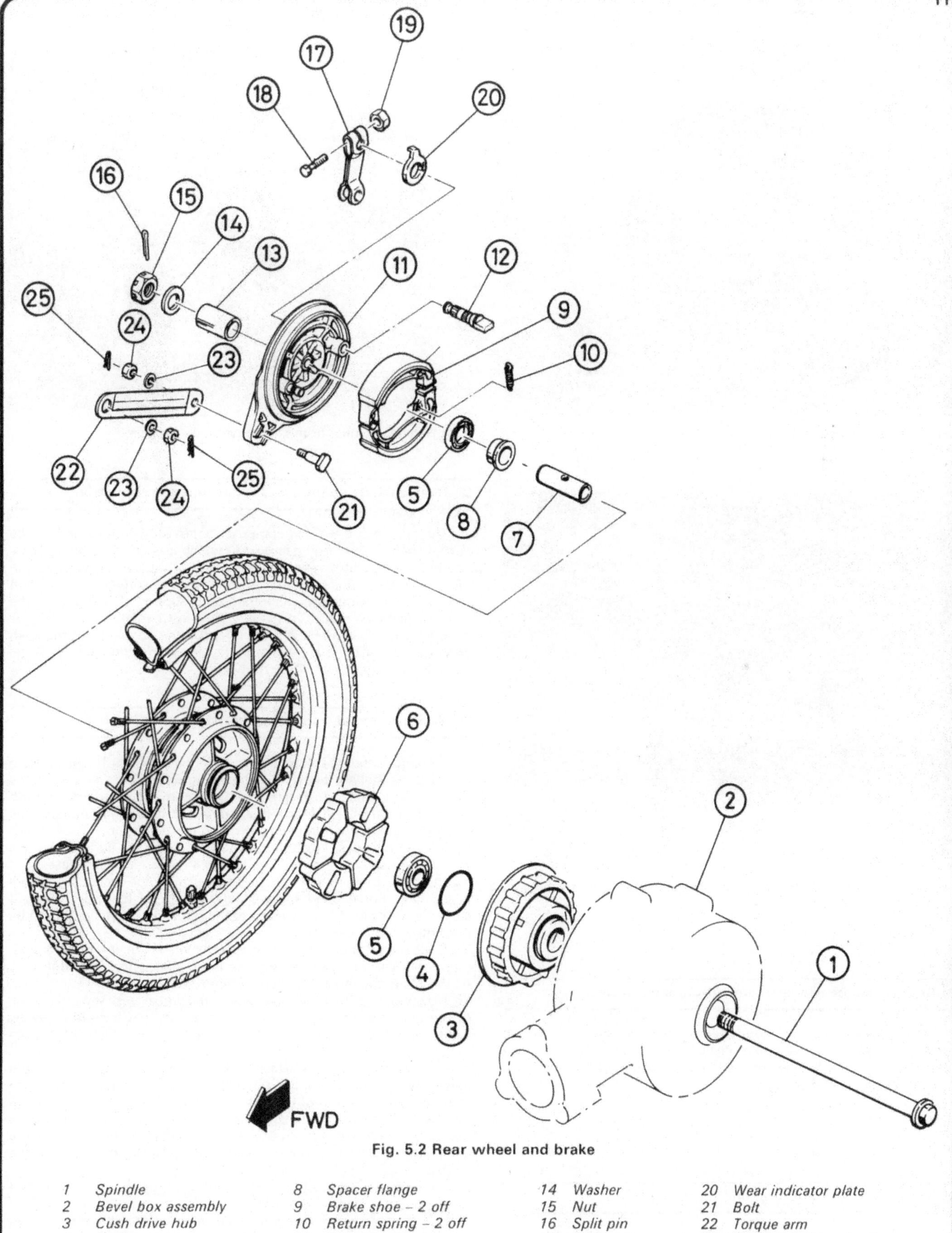

Fig. 5.2 Rear wheel and brake

1	*Spindle*	8	*Spacer flange*	14	*Washer*	20	*Wear indicator plate*
2	*Bevel box assembly*	9	*Brake shoe – 2 off*	15	*Nut*	21	*Bolt*
3	*Cush drive hub*	10	*Return spring – 2 off*	16	*Split pin*	22	*Torque arm*
4	*O-ring*	11	*Brake backplate*	17	*Brake arm*	23	*Spring washer – 2 off*
5	*Bearing – 2 off*	12	*Brake cam*	18	*Pinch bolt*	24	*Nut – 2 off*
6	*Cush drive rubber*	13	*Spacer*	19	*Nut*	25	*R-pin – 2 off*
7	*Spacer*						

6.2c ... then disengage and remove the wheel

6.2d Remove spindle from cush drive hub, if required

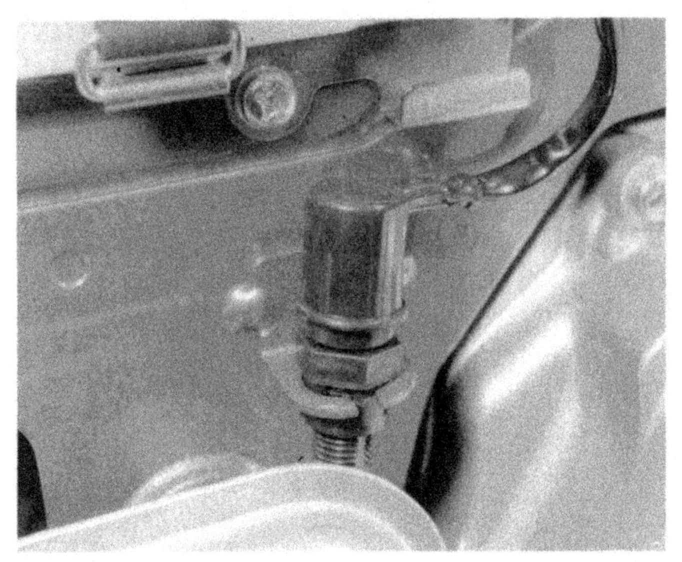

6.2e Check that the brake and switch are adjusted correctly during reassembly

7 Rear wheel bearings: examination and renewal

The rear wheel bearing arrangement is generally similar to that of the front wheel, and bearing examination and renewal can be tackled in the same way as described in Section 4. For details of the cush drive hub and bearing, see Section 9.

8 Rear brake: examination and renovation

Remove the rear wheel and disconnect the brake and torque arm as described in Section 6. Examination and renovation of the rear brake components is essentially the same as for the front brake and the details given in Section 5 can be applied. Note that unlike the front brake there is no return spring fitted to the operating arm.

9 Rear cush drive unit: examination and renovation

1 The rear cush drive unit comprises a cast alloy hub carried on the output shaft of the bevel gearbox, to which it is attached by a circlip. The hub incorporates vanes which engage in the cush drive rubber in the rear wheel hub, providing a method of absorbing transmission shocks. The bearings on which the hub is supported form part of the bevel gearbox assembly. This is described in Chapter 4.

2 Before the rear wheel is refitted, check the condition of the cush drive rubbers. If there are signs of cracking or deterioration, or if the assembly is sloppy, renew the rubbers.

10 Tyres: removal, repair and refitting

1 To remove the tyre from either wheel, first detach the wheel from the machine. Deflate the tyre by removing the valve core, and when the tyre is fully deflated, push the bead away from the wheel rim on both sides so that the bead enters the centre well of the rim. Remove the locking ring and push the tyre valve into the tyre itself.

2 Insert a tyre lever close to the valve and lever the edge of the tyre over the outside of the rim. Very little force should be necessary; if resistance is encountered it is probably due to the fact that the tyre beads have not entered the well of the rim, all the way round. If aluminium rims are fitted, damage to the soft alloy by tyre levers can be prevented by the use of plastic rim protectors.

3 Once the tyre has been edged over the wheel rim, it is easy to work round the wheel rim, so that the tyre is completely free from one side. At this stage the inner tube can be removed.

4 Now working from the other side of the wheel, ease the other edge of the tyre over the outside of the wheel rim that is furthest away. Continue to work around the rim until the tyre is completely free from the rim.

5 If a puncture has necessitated the removal of the tyre, reinflate the inner tube and immerse it in a bowl of water to trace the source of the leak. Mark the position of the leak, and deflate the tube. Dry the tube, and clean the area around the puncture with a petrol soaked rag. When the surface has dried, apply rubber solution and allow this to dry before removing the backing from the patch, and applying the patch to the surface.

6 It is best to use a patch of the self-vulcanizing type, which will form a permanent repair. Note that it may be necessary to remove a protective covering from the top surface of the patch after it has sealed into position. Inner tubes made from a special synthetic rubber may require a special type of patch and adhesive, if a satisfactory bond is to be achieved.

7 Before replacing the tyre, check the inside to make sure that the article that caused the puncture is not still trapped inside the tyre.

Tyre changing sequence — tubed tyres

Deflate tyre. After pushing tyre beads away from rim flanges push tyre bead into well of rim at point opposite valve. Insert tyre lever adjacent to valve and work bead over edge of rim.

Use two levers to work bead over edge of rim. Note use of rim protectors.

Remove inner tube from tyre.

When first bead is clear, remove tyre as shown.

When fitting, partially inflate inner tube and insert in tyre.

Work first bead over rim and feed valve through hole in rim. Partially screw on retaining nut to hold valve in place.

Check that inner tube is positioned correctly and work second bead over rim using tyre levers. Start at a point opposite valve.

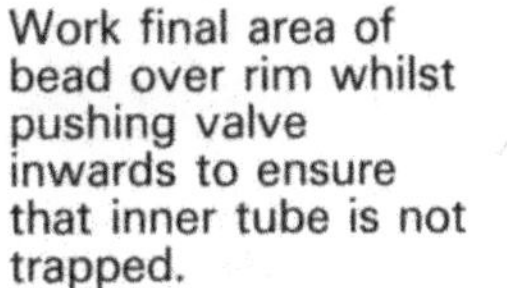

Work final area of bead over rim whilst pushing valve inwards to ensure that inner tube is not trapped.

Check the outside of the tyre, particularly the tread area, to make sure nothing is trapped that may cause a further puncture.

8 If the inner tube has been patched on a number of past occasions, or if there is a tear or large hole, it is preferable to discard it and fit a replacement. Sudden deflation may cause an accident, particularly if it occurs with the rear wheel.

9 To replace the tyre, inflate the inner tube for it just to assume a circular shape but only to that amount, and then push the tube into the tyre so that it is enclosed completely. Lay the tyre on the wheel at an angle, and insert the valve through the rim tape and the hole in the wheel rim. Attach the locking ring on the first few threads, sufficient to hold the valve captive in its correct location.

10 Starting at the point furthest from the valve, push the tyre bead over the edge of the wheel rim until it is located in the central well. Continue to work around the tyre in this fashion until the whole of one side of the tyre is on the rim. It may be necessary to use a tyre lever during the final stages.

11 Make sure there is no pull on the tyre valve and again commencing with the area furthest from the valve, ease the other bead of the tyre over the edge of the rim. Finish with the area close to the valve, pushing the valve up into the tyre until the locking ring touches the rim. This will ensure that the inner tube is not trapped when the last section of bead is edged over the rim with a tyre lever.

12 Check that the inner tube is not trapped at any point. Reinflate the inner tube, and check that the tyre is seating correctly around the wheel rim. There should be a thin rib moulded around the wall of the tyre on both sides, which should be an equal distance from the wheel rim at all points. If the tyre is unevenly located on the rim, try bouncing the wheel when the tyre is at the recommended pressure. It is probable that one of the beads has not pulled clear of the centre well.

13 Always run the tyres at the recommended pressures and never under or over inflate. The correct pressures are given in the Specifications Section of this Chapter.

14 Tyre replacement is aided by dusting the sidewalls, particularly in the vicinity of the beads, with a liberal coating of french chalk. Washing up liquid can also be used to good effect, but this has the disadvantage, where steel rims are used, of causing the inner surface of the wheel rim to rust.

15 Never replace the inner tube and tyre without the rim tape in position. If this precaution is overlooked there is a good chance of the ends of the spoke nipples chafing the inner tube and causing a crop of punctures.

16 Never fit a tyre that has a damaged tread or sidewalls. Apart from legal aspects, there is a very great risk of a blowout, which can have very serious consequences on a two wheeled vehicle.

17 Tyre valves rarely give trouble, but it is always advisable to check whether the valve itself is leaking before removing the tyre. Do not forget to fit the dust cap, which forms an effective extra seal.

11 Valve cores and caps

1 Valve cores seldom give trouble, but do not last indefinitely. Dirt under the seating will cause a puzzling 'slow-puncture'. Check that they are not leaking by applying spittle to the end of the valve and watching for air bubbles.

2 A valve cap is a safety device, and should always be fitted. Apart from keeping dirt out of the valve, it provides a second seal in case of valve failure, and may prevent an accident resulting from sudden deflation.

Chapter 6 Electrical system

Contents

Specifications

Electrical system

Voltage	6 volt
Earth (ground)	Negative (−)

Battery

Type	Lead-acid
Capacity	6 volt, 4 ampere-hours
Electrolyte specific gravity	1.260
Charging rate	0.4A maximum

Alternator

Make	Yamaha
Model	22K
Charging output:	
Minimum	0.8A at 3000 rpm
Maximum	2.0A at 8000 rpm
Charging coil resistance:	
T50 model	0.26 – 0.38 ohm at 20°C (68°F)
T80 model	0.32 ohm ± 10% at 20°C (68°F)
Lighting voltage:	
Minimum	6.2V at 3000 rpm
Maximum	8.0V at 8000 rpm
Lighting coil resistance:	
T50 model	0.17 – 0.25 ohm at 20°C (68°F)
T80 model	0.21 ohm ± 10% at 20°C (68°F)

Voltage regulator

No-load regulated voltage	7.0 – 7.6 volt

Rectifier

Withstand voltage	200 volt

Fuse

Rating	10A

Bulbs (all 6 volt)

Headlamp	25/25W
Tail/stop lamp	5/21W
Turn signal lamps:	
Early (35T) T80 model	15W
All T50 and later (2FL) T80 models	21W
Pilot lamp	3W
Instrument illuminating light	3W
Neutral indicator light	3W
High beam warning light – all T50 and later (2FL) T80 models	3W
Turn signal warning light – all T50 and later (2FL) T80 models	3W

Horn

Maximum amperage	1.5A

1 General description

The Yamaha T50 and 80 models are equipped with a 6 volt negative earth electrical system powered from the alternator (flywheel generator). The headlamp circuit runs direct from a lighting coil on the generator stator using the alternating current (ac) output without regulation or rectification. The rest of the system, including the turn signals, tail lamp, pilot lamp, brake lamp and instruments use direct current (dc). This is derived from a separate charging coil, and passes through an electronic regulator/rectifier unit before being passed to the battery.

The battery is rated at 6 volt 4 Ampere hour (Ah). It provides a reserve of power while the engine is idling, and thus not charging, and also allows a stable supply for circuits such as the turn signals, which would be adversely affected by a fluctuating supply.

2 Testing the electrical system: general

1 Simple continuity checks, for instance when testing switch units, wiring and connections, can be carried out using a battery and bulb arrangement to provide a test circuit. For most tests described in this chapter, however, a pocket multimeter should be considered essential. A basic multimeter capable of measuring volts and ohms can be bought for a very reasonable sum and will prove an invaluable tool. Note that separate volt and ohm meters may be used in place of the multimeter, provided those with the correct operating ranges are available. In addition, if the generator output is to be checked, an ammeter of 0-5 amperes range will be required.

2 Care must be taken when performing any electrical test, because some of the electrical components can be damaged if they are incorrectly connected or inadvertently shorted to earth. This is particularly so in the case of electronic components. Instructions regarding meter probe connections are given for each test, and these should be read carefully to preclude accidental damage occurring.

3 Where test equipment is not available, or the owner feels unsure of the procedure described, it is strongly recommended that professional assistance is sought. Errors made through carelessness or lack of experience can so easily lead to damage and need for expensive replacement parts.

4 For some of the tests, it may prove helpful to remove the legshield assembly. This is described in Chapter 4. The remaining major electrical components are housed below the right-hand side panel, around the battery holder, whilst the regulator/rectifier and the ignition (CDI) are mounted below the left-hand side panel. The switch connectors and associated wiring can be reached after the headlamp unit has been removed to provide access.

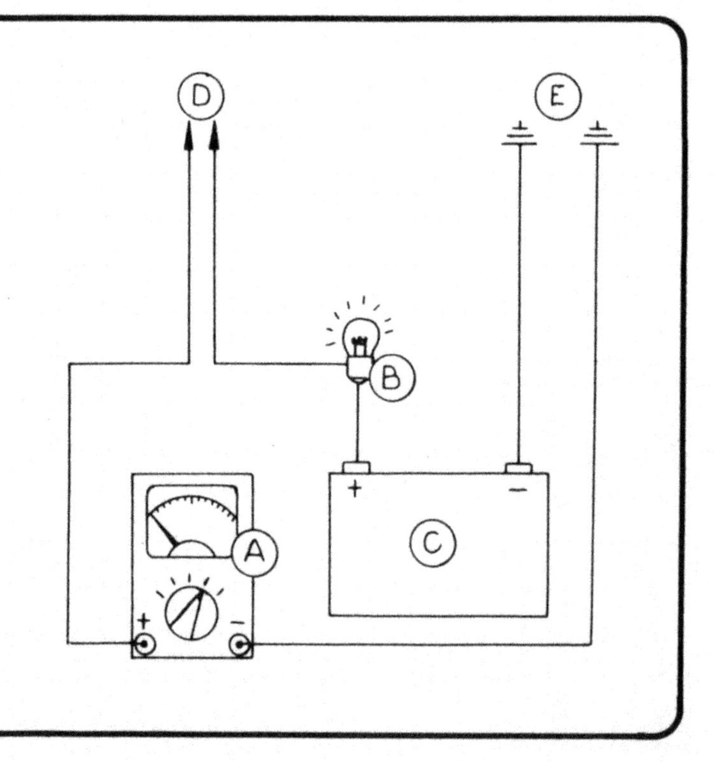

Fig. 6.1 Simple apparatus for testing the electrical system

A Multimeter
B Bulb
C Battery
D Positive probe
E Negative probe

3 Wiring: examination and layout

1 The wiring harness is colour-coded and will correspond with the wiring diagram at the end of this Chapter. When socket connectors are used, they are designed so that reconnection can be made in the correct position only.
2 Visual inspection will usually show whether there are any breaks or frayed outer coverings which will give rise to short circuits. Occasionally a wire may become trapped between two components, breaking the inner core but leaving the more resilient outer cover intact. This can give rise to mysterious intermittent or total circuit failure. Another source of trouble may be the snap connectors and sockets, where the connector has not been pushed fully home in the outer housing, or where corrosion has occurred.
3 Intermittent short circuits can often be traced to a chafed wire that passes through or is close to a metal component such as a frame member. Avoid tight bends in the lead or situations where a lead can become trapped between casings.

4 Battery: charging

1 During the winter months, or where numerous short journeys are undertaken, it is likely that the battery will require external charging from time to time. This should be done with the battery removed from the machine. It is strongly recommended that a small trickle charger of the type designed specifically for use on motorcycle batteries is used. The more common car-type chargers are voltage controlled, which means that it is not possible to regulate the charging current. It is likely that the charging current from equipment of this type will be much too high for small batteries, and there is a real risk of damage.
2 It should be noted in particular that the small size of the battery imposes limits on the rate at which it can safely be recharged. If too high a charge rate is applied, there is a small but real risk of the pressure bursting the battery, and the results of this can be imagined. Also, a high charge rate may overheat the battery, warping the plates and shortening its life significantly. The normal charge rate is 1/10th of the rated capacity of the battery, in this case 0.4 amps. A fully discharged battery will require about 5 hours of charging at this rate. In an emergency, a slightly higher charge rate (0.8 amp maximum) may be applied for short periods. Be warned that this is for emergencies only; it will shorten the life of the battery and must not be done often. If using the 'fast charge' rate, check the battery case frequently and stop charging **at once** if it gets more than slightly warm.
3 The battery is located behind the right-hand side panel and must be disconnected (negative lead first) and removed before charging commences. Loosen all of the cell caps and make sure that the breather is unobstructed. Never smoke near the battery or expose it to possible sources of fire. The gases released during charging are potentially explosive. Let the battery stand for some time after charging, and check the electrolyte level before it is refitted in the holder.

5 Charging system: checking the output

Warning: *When making any test on the electrical system with the engine running, make sure that the rear wheel is clear to rotate without touching the ground or other objects.*
1 This test requires the battery to be in good condition and fully charged – if in any doubt refer to Section 4. Also needed are an ammeter, or a multimeter capable of reading up to 5 amps; note that most small multimeters can only cope with milliamp currents – on no account use this scale or the meter will be damaged. If you do not have access to a suitable meter or are unsure of how to use it, the test should be performed by an authorized Yamaha dealer. Run the engine until normal operating temperature has been reached. Stop the engine and remove the right-hand side panel.
2 Open the fuse holder next to the battery and disconnect and separate the battery lead at the fuse terminal. Connect the ammeter between the lead and the fuse terminal. The red (+) meter probe should be connected to the fuse side of the holder, and the black (–) probe to the battery side.
3 Start the engine and observe the meter, while gradually increasing the engine speed. For accuracy, a test tachometer should be used to monitor engine speed, but in practice a useful indication of the charging output can be gained without this refinement. At 3000 rpm, the system should be producing a minimum of 0.8 amps, whilst at 8000 rpm, a reading of 2.0 amps maximum should be indicated. If the readings are other than specified, check first for loose or broken connections. If the ammeter shows a discharge, or if the readings are abnormally high or low, a faulty regulator/rectifier is indicated. Check this by measuring the regulator/rectifier resistances, or better still, by fitting a new unit. If the fault persists, check the condition of the alternator windings.

6 Alternator coils: testing

1 If it is suspected that a fault lies in the lighting or charging coils, this can be checked by using a multimeter to test the coil resistances. Start by removing the left-hand side panel to gain access to the various wiring connectors.
2 Trace the alternator wiring back to the connectors, and separate them. Using a multimeter set to the appropriate resistance scale, take a reading between the white and black wires. Compare the reading with the specified resistance of the charging coil in the Specifications section of this Chapter. Repeat the test between the yellow/red and black wires to measure the resistance of the lighting coil, again comparing the reading with that given in the Specifications.
3 If the test shows a reading of zero ohms, the coil insulation has broken down, allowing the coil to short to earth. If on the other hand, a reading of infinite resistance is shown, this indicates that the coil windings have broken. Either fault will require the renewal of the affected coil, but before ordering a new item it is worth having your findings confirmed by an expert.
4 The only real alternatives to renewal are to consult an auto-electrical specialist, who may be able to rewind the faulty coil, or to buy a second-hand stator from a motorcycle breaker. Removal and refitting of the alternator is described in the relevant Sections of Chapter 1.

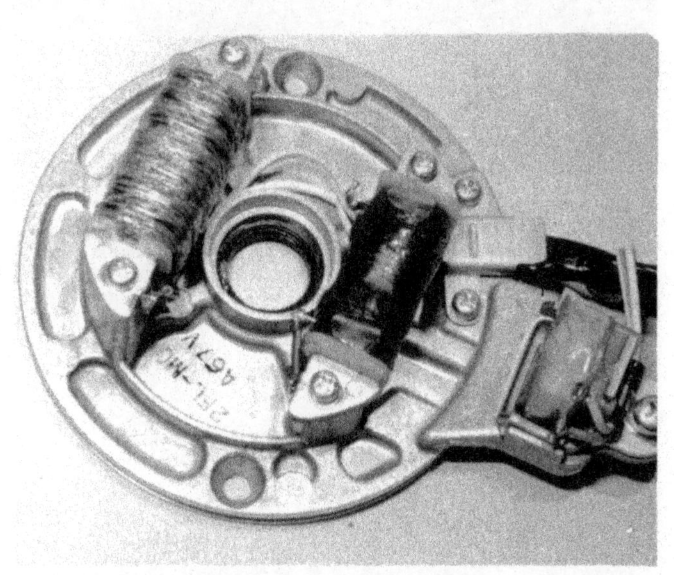

6.3 The alternator coils can be renewed individually, but must be soldered to the output wiring

7 Regulator/rectifier: testing

1 The regulator/rectifier takes the form of a sealed, finned alloy unit bolted to the frame behind the left-hand side panel. The unit rectifies the output of the charging coil, and also controls the charge rate applied to the battery. It is a fairly robust unit, but may be damaged by accidental short circuits, or in time, by mechanical damage from road shocks. This makes it a prime suspect in the event of a charging system fault.
2 Unplug the wiring connector from the underside of the unit then, using the accompanying table, check for continuity between the various pairs of terminals. If any one reading conflicts with the expected result, the unit can be assumed to have failed internally and must be renewed. It may be worth having your findings confirmed by

an authorized Yamaha dealer, or to check the unit by substitution before purchasing a replacement. Note that the unit is of sealed construction and cannot be repaired.

Meter Range	(+) probe connection	(−) probe connection	Continuity (Y or N)	Test current direction
R x 1	R	W	YES	Normal
R x 1	R	Y/R	YES	Normal
R x 1	W	Y/R	YES	Normal
R x 1	Y/R	R	NO	Reversed
R x 1	Y/R	W	NO	Reversed
R x 1	W	R	NO	Reversed

7.1 The regulator/rectifier is bolted to the frame, behind the left-hand side panel

8 Fuse: location and renewal

The electrical system is protected by a 10 amp fuse in an in-line holder in the battery positive lead. Access to the fuse is gained after removing the right-hand side panel. A spare fuse is housed within the holder. If the spare is used, be sure to obtain a new replacement of the correct rating at the earliest possible opportunity. In an emergency, a blown fuse can be bypassed by wrapping it in metal foil. Note that this should be done only if unavoidable, and bear in mind that the electrical system is left unprotected. Check and rectify any suspected fault before attempting to bypass the fuse.

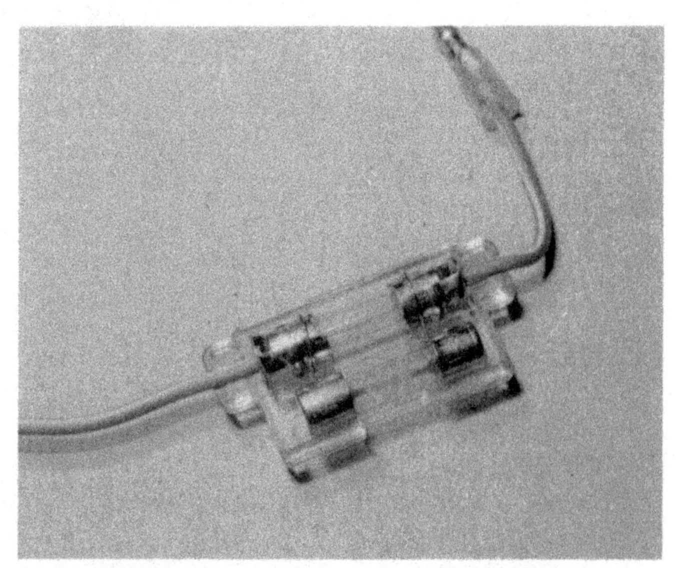

8.1 Fuseholder houses spare fuse – renew it if it gets used

9 Switches: examination and testing

1 The function of each of the various switches can be easily tested using a multimeter or a battery/bulb arrangement to check continuity in each of the switch positions. The switch harnesses are each routed down the side of the handlebar nacelle, and access to them can be gained after the headlamp unit has been removed. The headlamp unit is secured to the nacelle by two screws.

2 Referring to the wiring diagram at the end of this Chapter, you will note that each switch is represented in diagrammatic form, showing the wiring colours and which of these should be connected in each of the switch positions. Identify the connector belonging to the switch in question and unplug it. Connect the multimeter or battery/bulb arrangement across each pair of leads to be tested and check that it functions as shown on the switch diagram when the switch is operated. If a fault is indicated, or if operation is erratic, the switch must either be repaired or renewed as described below.

Ignition switch

3 The ignition switch is mounted in the instrument panel, the assembly being held in place by two screws. To remove the switch it will first be necessary to remove the speedometer head as described in Chapter 4. Once the speedometer has been removed from the nacelle, remove the two screws and lift the switch away.

4 If the operation of the switch has been affected by corrosion or the ingress of water, try soaking it in a maintenance fluid like WD40 and operating the switch repeatedly. If this fails to effect a cure, the switch will have to be renewed. If a new switch is fitted, remember that the new key will not necessarily fit the remaining locks on the machine, so either these will have to be renewed at the same time, or two keys will have to be carried.

Handlebar switches

5 The various handlebar switches are arranged at each end of the handlebar nacelle, access to them being gained by removing the screws which hold the switch halves together. If the operation of the switch has been affected by corrosion or the ingress of water, try soaking it in a maintenance fluid like WD40 and operating the switch repeatedly. This is quite often successful, the back of each switch being so designed that the fluid will penetrate readily. If this fails to effect a cure, the switch will have to be renewed. Before ordering a replacement, however, you may wish to attempt to dismantle and physically clean the switch terminals. The method of doing so is self-evident, but be warned that the switch may tend to fly apart when disturbed! It is suggested that the switch is removed and the work carried out on a clean bench so that small parts such as springs are not mislaid. If the attempt fails, and the switch was defective anyway, nothing will have been lost.

Brake lamp switches

6 The front brake lamp switch is a small plunger-type unit mounted on the brake lever stock. The switch is visible on the underside of the lever assembly. If the operation of the switch has been affected by corrosion or the ingress of water, try soaking it in a maintenance fluid like WD40 and operating the switch repeatedly. If this fails to restore operation, renew the switch; dismantling is not possible.

7 The rear brake switch is mounted near the brake pedal, and is operated by a small tension spring. The switch itself is sealed, and thus cannot be repaired. It is, however, adjustable. The switch body can be moved up or down in relation to its bracket after the locknuts have been slackened. The switch should be set to come on just after the pedal has been depressed.

Neutral switch

8 The neutral switch consists of a simple contact arrangement operated by a small cam on the gearbox selector drum. Faults are normally confined to a blown warning bulb or a broken or disconnected lead to the switch. If necessary, however, the switch can be unscrewed for examination after the gearchange pedal and the crankcase left-hand outer cover have been removed.

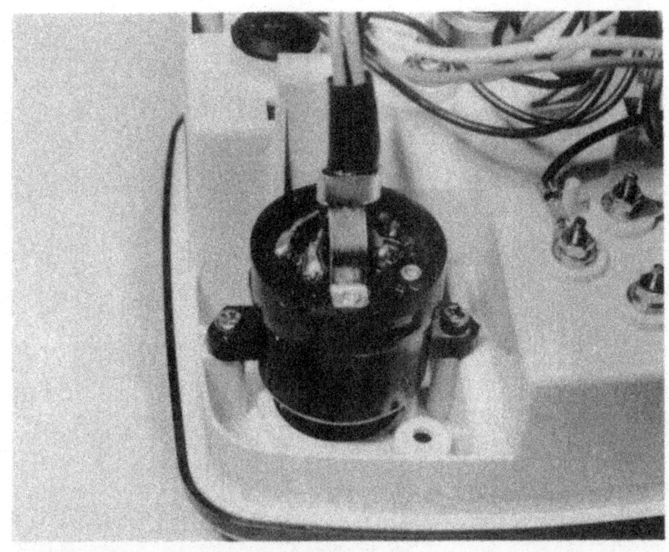

9.3 The ignition switch is screwed to the speedometer head

9.5a The right-hand handlebar switch cluster ...

9.5b ... and its left-hand counterpart

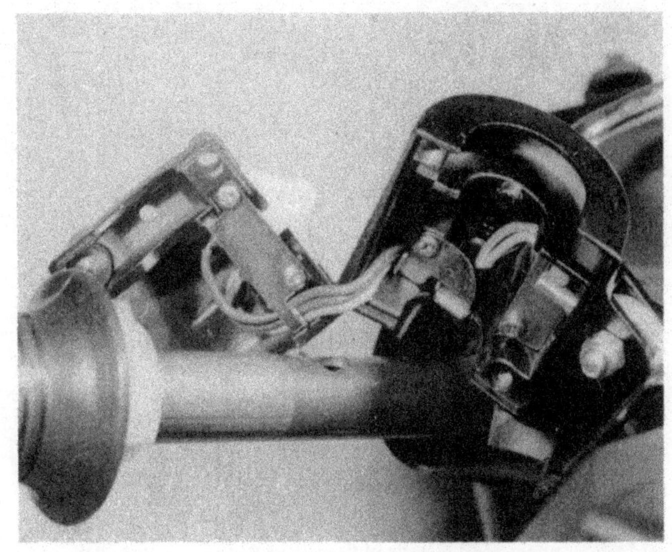

9.5c Switch halves can be separated for examination

10 Fuel gauge and sensor: removal and testing

1 The fuel gauge sensor is mounted in the top of the fuel tank and can be removed for testing after the seat has been opened. Slide back the plastic cover which covers the terminals on the unit to reveal the retaining screws. Remove the four screws to free the sensor unit. The sensor itself can now be lifted out of the tank, noting that it must be manoeuvred very carefully to avoid damage to the float or float arm. As the unit comes clear of the tank, pull the wiring out of the body pressing until the connector can be reached and separate the unit from the wiring harness.

2 The sensor consists of a float-operated variable resistance, and this can be checked by measuring the resistance of the unit at various float positions. Connect a multimeter across the terminals, setting it to the resistance (ohms) x 1 range. The resistance with the float fully raised (full position) should be 2 – 12 ohms, whilst at the fully lowered position (empty) it should read 87.5 – 97.5 ohms. The meter needle should move smoothly and progressively between the two readings as the float arm is raised and lowered. If the readings are significantly outside those given, or the resistance is erratic, the problem is usually corrosion of the resistance windings or the wiper arm.

3 Although a unit showing these symptoms is theoretically in need of renewal, it is possible to remove the pressed steel cover to gain access to the windings and wiper arm, or moving contact. The electrical contact surfaces can be cleaned using very fine abrasive paper or similar; an ink eraser is very good for this purpose and will not cause severe scratching. Finally, spray the windings with WD40 or similar, then repeat the resistance check to see if normal operation has been restored.

4 If the sensor operates normally, but there is still a fault, check that the electrical system in general functions normally by operating the turn signals, then connect the sensor leads without refitting the unit in the tank. Turn on the ignition switch and operate the float manually. If the gauge fails to respond, check back through the wiring to eliminate breaks or short circuits. If this does not resolve the problem, renew the gauge mechanism. The gauge is an integral part of the speedometer head, and this means that the complete assembly must be renewed if the gauge proves to be faulty.

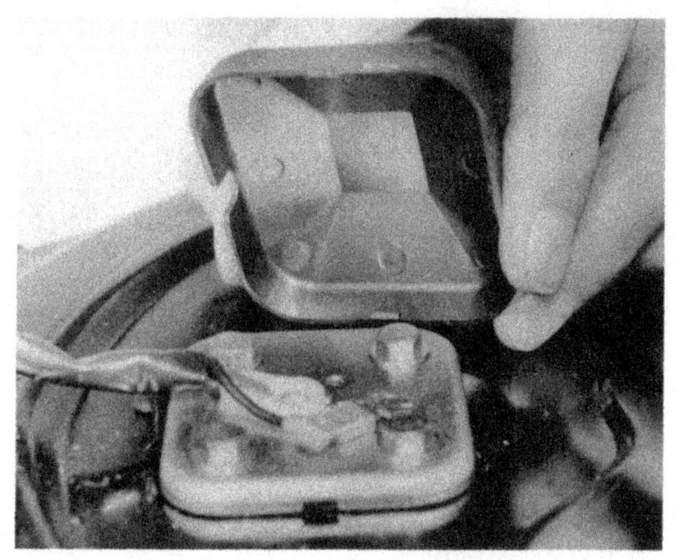

10.1a Fuel tank sender unit is housed below plastic cover

10.1b Be careful not to bend the float arm during removal

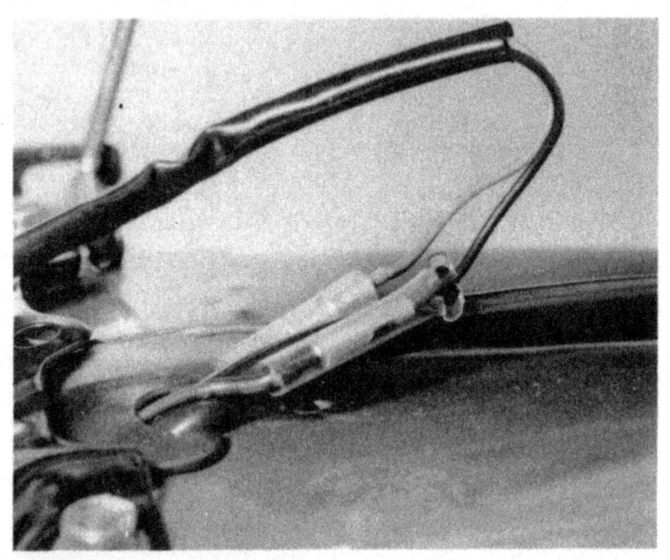

10.1c Reconnect sender leads and push back through grommet

11 Instrument panel warning bulbs: renewal

Access to the instrument panel bulbs requires the removal of the headlamp unit from the handlebar nacelle. The unit is secured by two screws and can be lifted away once these have been removed and the bulb wiring disconnected. Each of the bulbholders is a push-fit in the back of the instrument panel casing. Identify the bulb in question and withdraw its bulbholder. The bulb is a bayonet fit in its holder. When fitting a replacement bulb, make sure that one of the correct wattage is used.

11.1a Rear of speedometer showing bulbholders and wiring

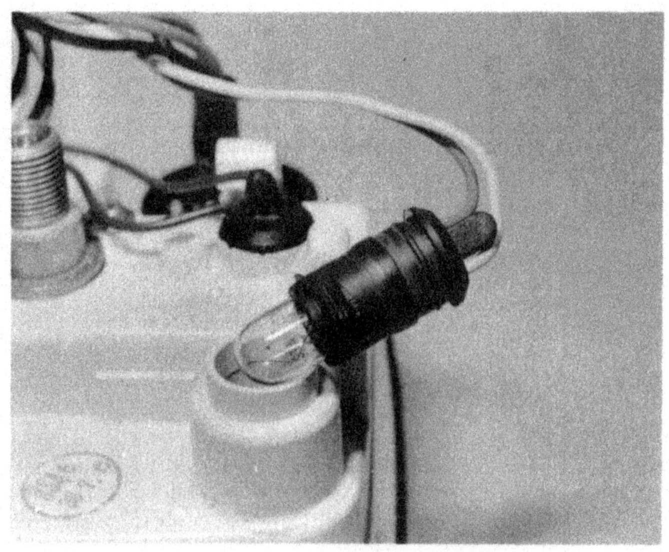

11.1b Bulbholders are a push fit in the speedometer

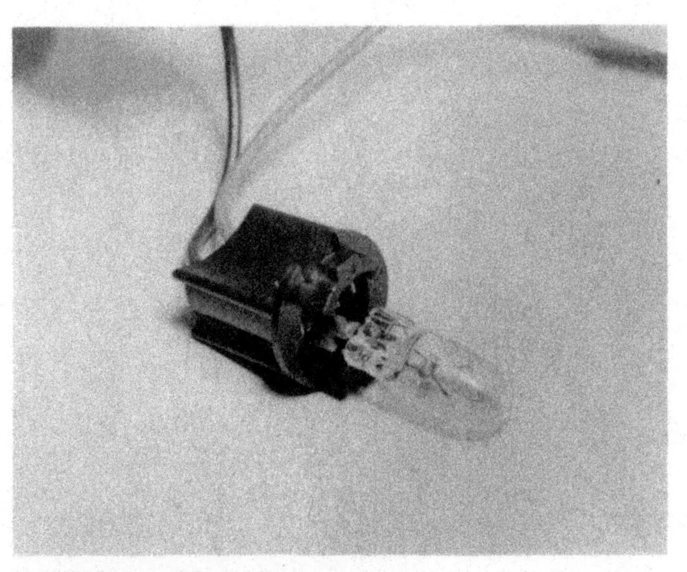

11.1c Bulbs are of the capless type

12 Headlamp: circuit output test, bulb renewal and beam alignment

1 Access to the headlamp and pilot lamp bulbs, or to the wiring connector for lighting circuit test purposes, requires the removal of the unit from the handlebar nacelle. The unit is retained by a screw on each side and may be lifted away once these have been released. Disconnect the wiring connector and place the unit to one side.

2 If the operation of the lighting system is suspect, the lighting output may be checked using a multimeter set on the ac 20V range. Connect the meter (+) probe to the yellow lead to the headlamp, and the (−) probe to the black lead. Turn the lighting switch on and select main beam, then start the engine. At 3000 rpm, a reading of at least 6.2 volts should be shown, whilst at 8000 rpm, no more than 8.0 volts should be produced. In the absence of a test tachometer, the engine speed will have to be approximated, but even without the meter a rough indication can be gained. The exact reading is nor crucial, but if the output varies significantly from the above figures, check the regulator/rectifier unit and the alternator coil resistances as described earlier in this Chapter.

3 To gain access to the headlamp bulbs, remove the unit from its nacelle as described above. To remove the headlamp bulb, depress the bulbholder and turn it anticlockwise. Lift the bulbholder clear to reveal the back of the bulb. The new bulb can be fitted by reversing the above sequence. The pilot lamp bulb is a bayonet fitting in its holder, this being a push-fit in the back of the reflector.

4 In the UK the headlamp beam must be set as follows: With the machine standing on its wheels on level ground and the rider (and pillion passenger if one is regularly carried) seated normally, check that the dip beam centre (that part which shows the greatest intensity of light) is just below (at least 0.5°) the level of the headlamp centre when it is shining on a wall 12.5 ft (3.84 m) away. Note that the tyre pressures and rear suspension settings must be set to suit the load applied. If the headlamp requires adjustment, this can be carried out by moving the adjuster screw to the required position. The screw is located on the edge of the headlamp rim. Note that most countries have laws governing the setting of beam alignment, and adjustments must be made with this in mind.

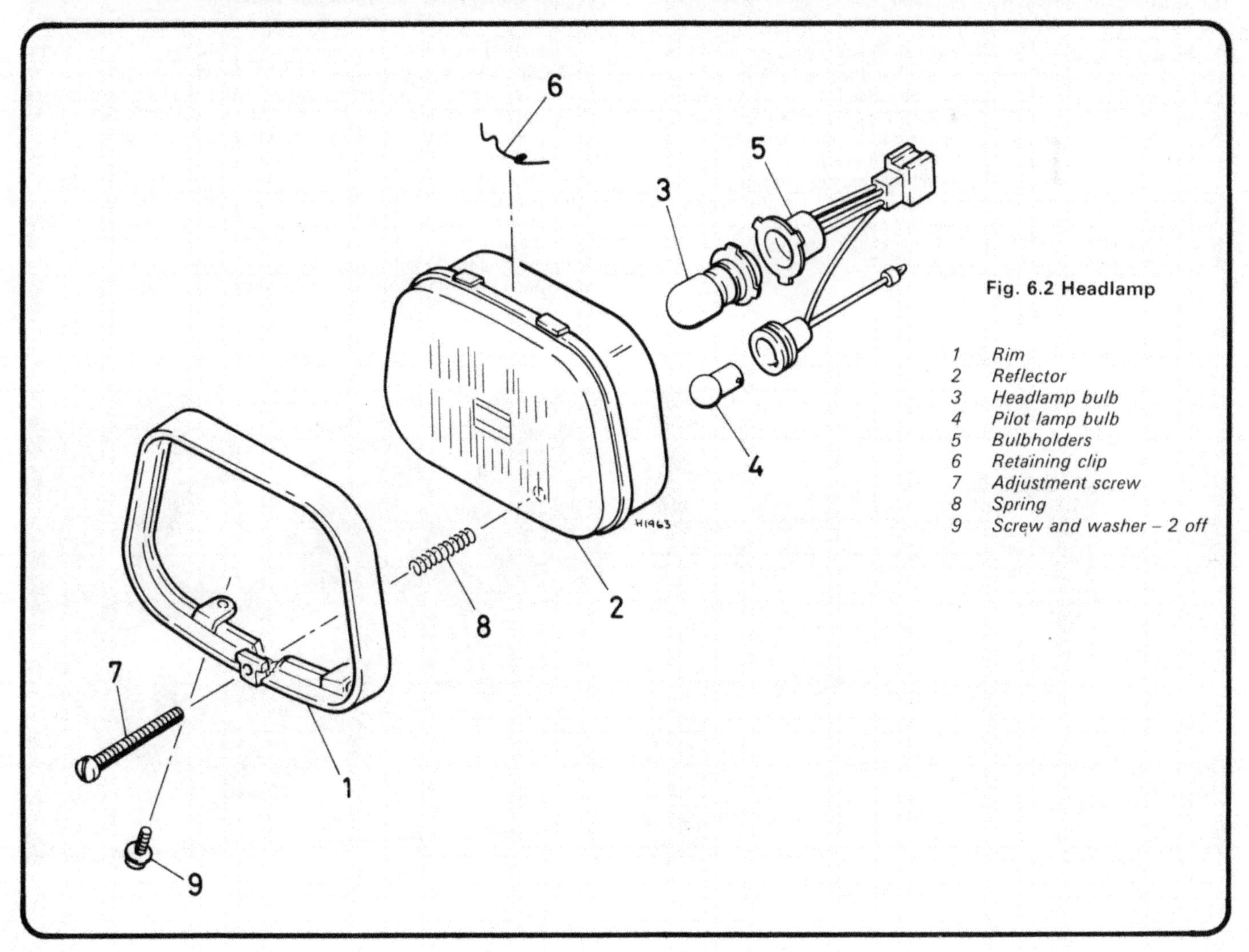

Fig. 6.2 Headlamp

1 *Rim*
2 *Reflector*
3 *Headlamp bulb*
4 *Pilot lamp bulb*
5 *Bulbholders*
6 *Retaining clip*
7 *Adjustment screw*
8 *Spring*
9 *Screw and washer – 2 off*

12.3a Twist bulbholder to release it from headlamp ...

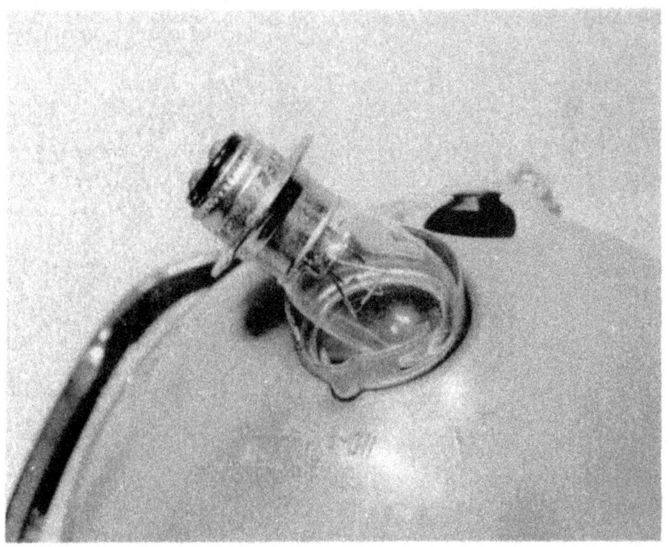
12.3b ... and lift bulb away

13 Turn signal lamps: bulb renewal

1 Bulb failure is by far the most common cause of turn signal problems, and is characterised by one lamp failing to flash, while the other flashes dimly and rapidly. The other pair of lamps will be unaffected. Having identified the bulb at fault, proceed as follows.
2 In the case of both the front and rear turn signal lamps, remove the two screws which retain the lens and lift it away.
3 In each case, fit a new bulb, and check that the turn signals operate normally before refitting the lens. If the fault persists or the bulb proves not to have failed, check for a wiring fault in the turn signal circuit. Note that failure of one side of the circuit only is not normally attributable to the turn signal relay.

14 Turn signal relay: location and renewal

1 If there is a fault in the turn signal circuit, always check the bulbs (see Section 13) and the wiring and switch before turning attention to the relay. It is worth noting that a partially discharged battery will affect the operation of the turn signal system. Note also that a fault in the relay will affect all four lamps; not just one lamp or pair of lamps.
2 If the relay develops a fault it will be necessary to renew it. Access to the relay requires the removal of the headlamp unit. The relay is a small rectangular black unit plugged into the wiring harness.
3 Renewal is a simple matter of unplugging the defective relay and plugging in a new one. Check that the system functions normally, then refit the headlamp. If there is a problem with the rate at which the turn signal lamps flash, check that the bulb wattages are correct.

13.2 Release turn signal lamp lens to gain access to bulb

14.2 Turn signal relay plugs into wiring inside headlamp

16 Horn: location and renewal

The horn is mounted on the front of the machine, beneath the legshields. Check the horn by removing it and connecting a 6 volt battery to its terminals. If the horn fails to sound, it must be renewed; repair is not practicable. If the horn operates under test, but not on the machine, check the wiring and switch operation as described earlier in this Chapter.

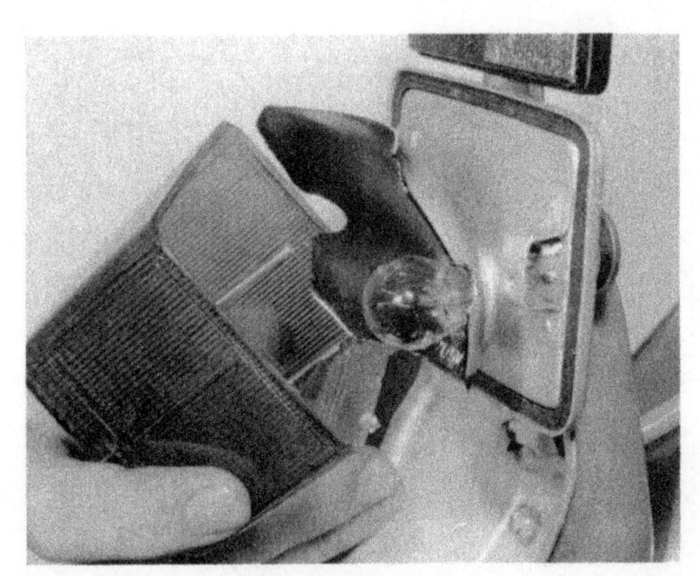

15.1 Release stop/tail lamp lens to gain access to bulb

15 Stop/tail lamp: bulb renewal

Access to the stop/tail bulb is gained by removing the two screws which retain the lens unit and lifting it away. When renewing the bulb, check that it is of the correct wattage, and note that the pins are offset to prevent it from being fitted incorrectly. If the brake or tail lamps fail to operate despite the bulb being intact, trace back through the wiring, checking for a break or short, and check the operation of the lighting and brake lamp switches.

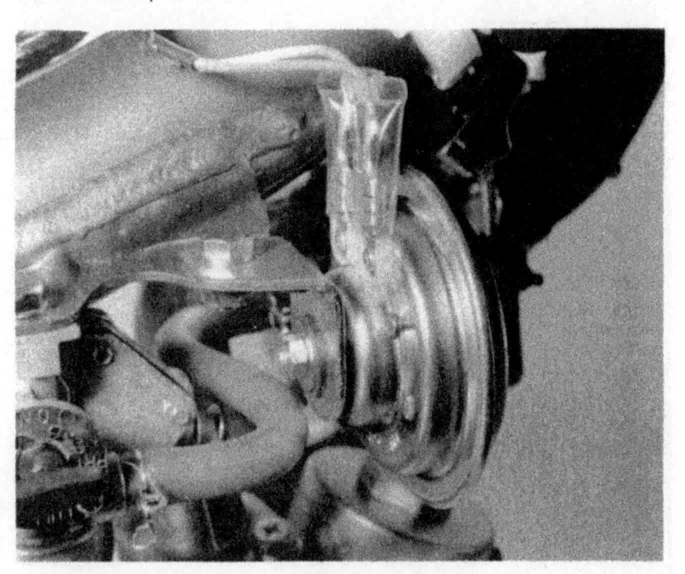

16.1 Horn is located on frame below legshield

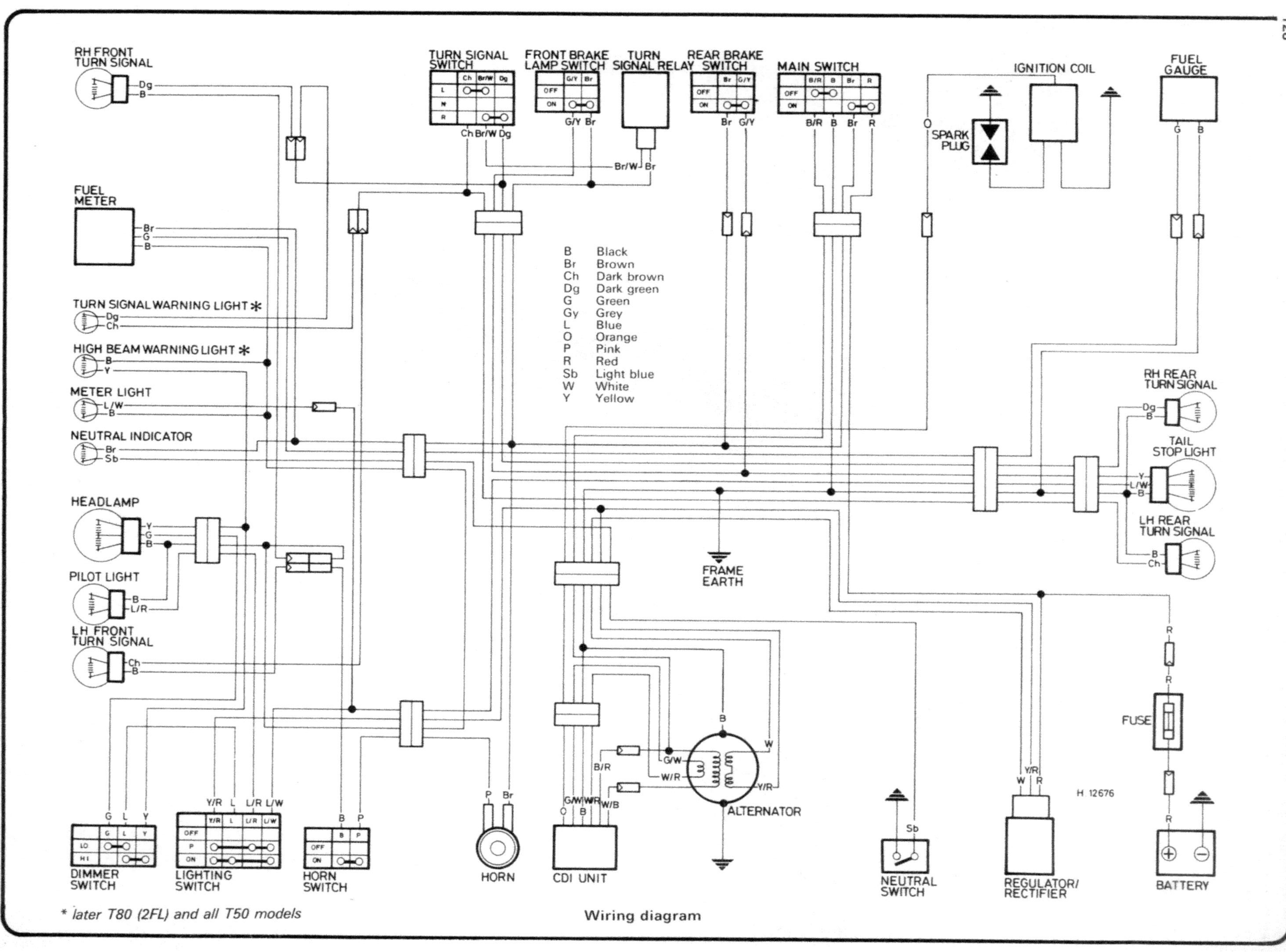

* later T80 (2FL) and all T50 models

Wiring diagram

Conversion factors

Length (distance)						
Inches (in)	X	25.4	= Millimetres (mm)	X	0.0394	= Inches (in)
Feet (ft)	X	0.305	= Metres (m)	X	3.281	= Feet (ft)
Miles	X	1.609	= Kilometres (km)	X	0.621	= Miles
Volume (capacity)						
Cubic inches (cu in; in^3)	X	16.387	= Cubic centimetres (cc; cm^3)	X	0.061	= Cubic inches (cu in; in^3)
Imperial pints (Imp pt)	X	0.568	= Litres (l)	X	1.76	= Imperial pints (Imp pt)
Imperial quarts (Imp qt)	X	1.137	= Litres (l)	X	0.88	= Imperial quarts (Imp qt)
Imperial quarts (Imp qt)	X	1.201	= US quarts (US qt)	X	0.833	= Imperial quarts (Imp qt)
US quarts (US qt)	X	0.946	= Litres (l)	X	1.057	= US quarts (US qt)
Imperial gallons (Imp gal)	X	4.546	= Litres (l)	X	0.22	= Imperial gallons (Imp gal)
Imperial gallons (Imp gal)	X	1.201	= US gallons (US gal)	X	0.833	= Imperial gallons (Imp gal)
US gallons (US gal)	X	3.785	= Litres (l)	X	0.264	= US gallons (US gal)
Mass (weight)						
Ounces (oz)	X	28.35	= Grams (g)	X	0.035	= Ounces (oz)
Pounds (lb)	X	0.454	= Kilograms (kg)	X	2.205	= Pounds (lb)
Force						
Ounces-force (ozf; oz)	X	0.278	= Newtons (N)	X	3.6	= Ounces-force (ozf; oz)
Pounds-force (lbf; lb)	X	4.448	= Newtons (N)	X	0.225	= Pounds-force (lbf; lb)
Newtons (N)	X	0.1	= Kilograms-force (kgf; kg)	X	9.81	= Newtons (N)
Pressure						
Pounds-force per square inch (psi; lbf/in^2; lb/in^2)	X	0.070	= Kilograms-force per square centimetre (kgf/cm^2; kg/cm^2)	X	14.223	= Pounds-force per square inch (psi; lbf/in^2; lb/in^2)
Pounds-force per square inch (psi; lbf/in^2; lb/in^2)	X	0.068	= Atmospheres (atm)	X	14.696	= Pounds-force per square inch (psi; lbf/in^2; lb/in^2)
Pounds-force per square inch (psi; lbf/in^2; lb/in^2)	X	0.069	= Bars	X	14.5	= Pounds-force per square inch (psi; lbf/in^2; lb/in^2)
Pounds-force per square inch (psi; lbf/in^2; lb/in^2)	X	6.895	= Kilopascals (kPa)	X	0.145	= Pounds-force per square inch (psi; lbf/in^2; lb/in^2)
Kilopascals (kPa)	X	0.01	= Kilograms-force per square centimetre (kgf/cm^2; kg/cm^2)	X	98.1	= Kilopascals (kPa)
Torque (moment of force)						
Pounds-force inches (lbf in; lb in)	X	1.152	= Kilograms-force centimetre (kgf cm; kg cm)	X	0.868	= Pounds-force inches (lbf in; lb in)
Pounds-force inches (lbf in; lb in)	X	0.113	= Newton metres (Nm)	X	8.85	= Pounds-force inches (lbf in; lb in)
Pounds-force inches (lbf in; lb in)	X	0.083	= Pounds-force feet (lbf ft; lb ft)	X	12	= Pounds-force inches (lbf in; lb in)
Pounds-force feet (lbf ft; lb ft)	X	0.138	= Kilograms-force metres (kgf m; kg m)	X	7.233	= Pounds-force feet (lbf ft; lb ft)
Pounds-force feet (lbf ft; lb ft)	X	1.356	= Newton metres (Nm)	X	0.738	= Pounds-force feet (lbf ft; lb ft)
Newton metres (Nm)	X	0.102	= Kilograms-force metres (kgf m; kg m)	X	9.804	= Newton metres (Nm)
Power						
Horsepower (hp)	X	745.7	= Watts (W)	X	0.0013	= Horsepower (hp)
Velocity (speed)						
Miles per hour (miles/hr; mph)	X	1.609	= Kilometres per hour (km/hr; kph)	X	0.621	= Miles per hour (miles/hr; mph)
Fuel consumption*						
Miles per gallon, Imperial (mpg)	X	0.354	= Kilometres per litre (km/l)	X	2.825	= Miles per gallon, Imperial (mpg)
Miles per gallon, US (mpg)	X	0.425	= Kilometres per litre (km/l)	X	2.352	= Miles per gallon, US (mpg)

Temperature

Degrees Fahrenheit = (°C x 1.8) + 32

Degrees Celsius (Degrees Centigrade; °C) = (°F - 32) x 0.56

**It is common practice to convert from miles per gallon (mpg) to litres/100 kilometres (l/100km), where mpg (Imperial) x l/100 km = 282 and mpg (US) x l/100 km = 235*

Index

O

P

R

S

T

V

W